I0020706

Building a Next-Gen SOC with IBM QRadar

Accelerate your security operations and detect
cyber threats effectively

Ashish M Kothekar

BIRMINGHAM—MUMBAI

Building a Next-Gen SOC with IBM QRadar

Copyright © 2023 Packt Publishing

All rights reserved. No part of this book may be reproduced, stored in a retrieval system, or transmitted in any form or by any means, without the prior written permission of the publisher, except in the case of brief quotations embedded in critical articles or reviews.

Every effort has been made in the preparation of this book to ensure the accuracy of the information presented. However, the information contained in this book is sold without warranty, either express or implied. Neither the author, nor Packt Publishing or its dealers and distributors, will be held liable for any damages caused or alleged to have been caused directly or indirectly by this book.

Packt Publishing has endeavored to provide trademark information about all of the companies and products mentioned in this book by the appropriate use of capitals. However, Packt Publishing cannot guarantee the accuracy of this information.

Group Product Manager: Pavan Ramchandani

Publishing Product Manager: Prachi Sawant

Senior Editor: Romy Dias

Technical Editor: Arjun Varma

Copy Editor: Safis Editing

Project Coordinator: Ashwin Kharwa

Proofreader: Safis Editing

Indexer: Pratik Shirodkar

Production Designer: Prashant Ghare

Marketing Coordinator: Marylou Dmello

First published: June 2023

Production reference: 1310523

Published by Packt Publishing Ltd.

Livery Place

35 Livery Street

Birmingham

B3 2PB, UK.

ISBN 978-1-80107-602-9

www.packtpub.com

I strongly believe in the power of manifestation. This book is the manifestation of my utmost effort to take up complex subjects, simplify them, and present them to a larger audience.

Foreword

IBM QRadar has been the leader in the security market segment for more than a decade. IBM QRadar is an aggregator that encompasses security products, ingests data from them, correlates the data, and uses artificial intelligence and machine learning to provide insights, alerts, and reports for practitioners and stakeholders. These include analysts, security administrators, government agencies, law enforcement, and executive management. This book gets into the foundational concepts on which QRadar is built, explaining the architecture, services, and different components. It details how QRadar can be used by stakeholders for different purposes, covering QRadar apps, which are customized solutions built by IBM for different security products. It ends with informative crafted chapters on the author's experience with QRadar and tips for getting the most out of QRadar for your organization's security.

I have known Ashish for over half a decade and closely worked with him, where he has been instrumental in designing different security solutions, including IBM QRadar and IBM Storage products. He has authored many IBM Redpapers in this area. This book provides an end-to-end view and knowledge of QRadar, including insights into its usage to secure your organization. I definitely recommend it for practitioners across industry verticals.

Sandeep Patil

IBM STSM (IBM Storage CTO Office), IBM Master Inventor, WW Prolific Inventor

Contributors

About the author

Ashish M Kothekar is currently working on the SWAT team for IBM Security products. He has more than 16 years of experience working with IBM. He is the **subject matter expert** (**SME**) for IBM products in the threat management segment, including IBM QRadar and Cloud Pak solutions for security. He is the author of many IBM Redpapers written on the collaboration of IBM Security and IBM Storage products. He is an avid tech blogger and writes on various security use cases.

I would like to take this opportunity to thank Yogesh Talekar, my manager, for his consent and for all his help in fast-tracking the formalities required to start this book. Yogesh is more than just a manager; he is also a friend, guide, and philosopher.

The technical reviewers—Boudhyayan Chakrabarty (Bob), Sam Yiu, and Ankit Rai—have helped me dive deep into the technical aspects of QRadar features. Special shout-out to Bob, who has been there through thick and thin in this journey.

My family has stood by me and been a source of encouragement throughout this journey. I would like to sincerely thank them. And to my late grandmother, without whom many things would not have been possible, including this book.

About the reviewers

Samson Yiu is a cybersecurity engineer whose qualifications include a degree in computer science; CISSP, IBM Security Certified Consultant, ITIL, MCSE, and postgraduate certificates in cybersecurity; and detailed knowledge of security architectures and best practices. He has 12 years of experience in the design, implementation, and support of solutions protecting networks, systems, and information assets for Fortune 500 companies throughout Asia. His major strength is in troubleshooting failed deployments and bad architectural implementations. He has served as the SME for QRadar certification exams and is currently engaged as an active NSW cybersecurity ambassador. This is the sixth book that he has been involved in.

I wish to thank my wife, Jessie, and my kids, Edison, Vivian, and Rochelle, for allowing me the time to improve myself to study and write. Cybersecurity as a job did not exist when I graduated from university so I wish those who have chosen this vocation the best of luck in the future as this is an awesome career with infinite learning.

Ankit Rai has more than 10 years' experience in cybersecurity and has assisted the CISOs, VPs, and heads of cybersecurity departments at various multinational corporations in procurement and doing PoC of the security tools, as well as performing end-to-end deployment, maintenance, and support for various cybersecurity solutions. Ankit Rai is a successful professional who built the first ever fully functional SOC for one of the largest small finance banks. He has a large community across social networking sites and loves to speak about cybersecurity in colleges and universities.

My heart goes out to my family, who made me the man I am today, and to my managers, who always promoted my urge to learn and my unquenchable curiosity. I'd also like to thank my life partner, Monika, who has been with me through thick and thin, night and day, always.

Lastly, I'd like to thank the community who welcomed me with open arms when I was still learning the ropes of information security. Thank you!

Boudhayan Chakrabarty (Bob) is an executive architect in the IBM Security Elite team specializing in threat management. With over 16 years of hands-on experience, Bob has also been a part of many international deployments of security compliance and security intelligence solutions, wherein he contributed starting from the RFP phase to the proof of concept phase to the ultimate deployment and training of the customer. He is a regular speaker on cybersecurity at different conferences and has also authored multiple books and publications. He has also run many enablement and training courses on security intelligence and compliance products. He is an SME for Security Intelligence and Compliance certifications.

I would like to take this opportunity and thank everyone who had supported me in this journey, especially all my mentors. Special mention to Yogesh Talekar who has been a source of inspiration and a guiding light for me always. He is the reason for whatever I have achieved so far. I would also like to thank my life Maitreyee who has always stood up for me and by me. She sustained me in ways I could never imagine. Thank you for everything.

Table of Contents

Part 2: QRadar Features and Deployment

6

QRadar Searches   71

7

QRadar Rules and Offenses   87

Part 3: Understanding QRadar Apps, Extensions, and Their Deployment

8

The Insider Threat – Detection and Mitigation   105

Preface

This book is a complete guide to planning, deploying, and managing a QRadar environment for any security operation center. It focuses on the intricacies of each component in QRadar and how they can be deployed to achieve the desired result. You will find the best practices to implement huge deployments in QRadar. This book describes how QRadar apps should be used as added features as well as how to develop customized QRadar apps.

Who this book is for

This book is for security analysts, system administrators, and security architects, as well as executive management to help them understand the concepts and features of QRadar. The book includes real-world examples that will help incident management teams handle security incidents and plan for cybersecurity attacks.

What this book covers

Chapter 1, *QRadar Components*, explains all the QRadar components, what the different QRadar services are, and which services run on which components. This chapter will help you understand how QRadar is designed and how different components provide different functionalities.

Chapter 2, *How QRadar Components Fit Together*, looks at the QRadar console, which is the central component around which other components fit together; depending on the requirement, other QRadar components can be added to the console. Also, we will explain in detail what different types of deployments exist – namely, all-in-one deployment and distributed deployment.

Chapter 3, *Managing QRadar Deployments*, deals with installing, upgrading, and scaling QRadar as and when required. We also discuss licensing requirements in QRadar.

Chapter 4, *Integrating Logs and Flows in QRadar*, discusses the practical aspects of ingesting data in QRadar. There are various ways in which different types of events and flow data are ingested, which are described in detail in this chapter.

Chapter 5, *Leaving No Data Behind*, explores how data is handled by QRadar. The majority of the shortcomings when working with QRadar occur while ingesting data. We will also discuss the DSM Editor, a tool to ingest data that is not supported out of the box.

Chapter 6, *QRadar Searches*, discusses how searches work and how they can be tuned in QRadar. SIEM is only as efficient as the searches performed on it. We will also discuss the different types of searches in QRadar and how data accumulation works in it.

Chapter 7, QRadar Rules and Offenses, delves into one of the most fundamental aspects of QRadar, which is rules and offenses. We will discuss the different types of rules, how to run rules for historical data called historical correlation, how offenses are generated, and finally, how to fine-tune and manage rules and offenses.

Chapter 8, The Insider Threat – Detection and Mitigation, examines how UBA can be used to detect an insider threat in your organization. IBM has a public portal where apps are published, which can be downloaded and installed on QRadar. Some of these apps are created by IBM, while other vendors have come up with apps for their own applications. IBM UBA is one such app developed by IBM for insider threat management.

Chapter 9, Integrating AI into Threat Management, discusses three QRadar apps – the QRadar Assistant app, QRadar Advisor for Watson, and QRadar Use Case Manager. We will also discuss the practical use of these apps.

Chapter 10, Re-Designing User Experience, explores how to use apps to improve the user experience. IBM QRadar needed an overhaul when it came to user experience. Hence, IBM devised apps such as IBM QRadar Pulse and IBM Analyst Workflow to change the way QRadar can be managed, which we will look at in this chapter.

Chapter 11, WinCollect – the Agent for Windows, focuses on how to install, manage, upgrade, and fine-tune Wincollect agents, one of many in-built features from IBM QRadar. Wincollect is an agent for the Windows operating system and collects events from Windows machines. It can also poll events from other Windows machines where it is not installed and send them to QRadar.

Chapter 12, Troubleshooting QRadar, examines the pain points and solutions to many of the issues in QRadar, based on years of experience working with it. There are tips and tricks as well as a list of frequently asked questions about QRadar. This chapter should help you become a pro user of QRadar.

To get the most out of this book

Software/hardware covered in the book	Operating system requirements
QRadar components and services	Windows, macOS, or Linux – for browsing and logging in to a command-line interface
Wincollect	RHEL – for QRadar

If you are using the digital version of this book, we advise you to type the code yourself or access the code from the book's GitHub repository (a link is available in the next section). Doing so will help you avoid any potential errors related to the copying and pasting of code.

Download the color images

We also provide a PDF file that has color images of the screenshots and diagrams used in this book. You can download it here: https://packt.link/PtEjQ.

Conventions used

There are a number of text conventions used throughout this book.

Code in text: Indicates code words in text, database table names, folder names, filenames, file extensions, pathnames, dummy URLs, user input, and Twitter handles. Here is an example: "The log source name as defined here would be cali_ips@9.9.9.9, where cali_ips could be the device type and the source address could be the source IP picked up from the event payload."

A block of code is set as follows:

```
html, body, #map {
  height: 100%;
  margin: 0;
  padding: 0
}
```

When we wish to draw your attention to a particular part of a code block, the relevant lines or items are set in bold:

```
[default]
exten => s,1,Dial(Zap/1|30)
exten => s,2,Voicemail(u100)
exten => s,102,Voicemail(b100)
exten => i,1,Voicemail(s0)
```

Any command-line input or output is written as follows:

```
$ mkdir css
$ cd css
```

Bold: Indicates a new term, an important word, or words that you see on screen. For instance, words in menus or dialog boxes appear in **bold**. Here is an example: "For this feature, you need to enable it in the **Configuration** tab of the DSM Editor."

> **Tips or important notes**
> Appear like this.

Get in touch

Feedback from our readers is always welcome.

General feedback: If you have questions about any aspect of this book, email us at customercare@ packtpub.com and mention the book title in the subject of your message.

Errata: Although we have taken every care to ensure the accuracy of our content, mistakes do happen. If you have found a mistake in this book, we would be grateful if you would report this to us. Please visit www.packtpub.com/support/errata and fill in the form.

Piracy: If you come across any illegal copies of our works in any form on the internet, we would be grateful if you would provide us with the location address or website name. Please contact us at copyright@packtpub.com with a link to the material.

If you are interested in becoming an author: If there is a topic that you have expertise in and you are interested in either writing or contributing to a book, please visit authors.packtpub.com.

Share your thoughts

Once you've read *Building a Next-Gen SOC with IBM QRadar*, we'd love to hear your thoughts! Scan the QR code below to go straight to the Amazon review page for this book and share your feedback.

https://packt.link/r/1801076022

Your review is important to us and the tech community and will help us make sure we're delivering excellent quality content.

Download a free PDF copy of this book

Thanks for purchasing this book!

Do you like to read on the go but are unable to carry your print books everywhere?

Is your eBook purchase not compatible with the device of your choice?

Don't worry, now with every Packt book you get a DRM-free PDF version of that book at no cost.

Read anywhere, any place, on any device. Search, copy, and paste code from your favorite technical books directly into your application.

The perks don't stop there, you can get exclusive access to discounts, newsletters, and great free content in your inbox daily

Follow these simple steps to get the benefits:

1. Scan the QR code or visit the link below

https://packt.link/free-ebook/9781801076029

2. Submit your proof of purchase
3. That's it! We'll send your free PDF and other benefits to your email directly

Part 1:
Understanding Different
QRadar Components and
Architecture

In this part, we work on the fundamentals of QRadar by discussing different types of components and how they fit together. We also consider the different types of deployments and how to manage, scale, and upgrade them.

This part has the following chapters:

QRadar Components

We live in a digital age in which the paradigms of security have changed. In the past, wars were fought on battlefields. Now, digital space is where the security of a nation-state, an enterprise, or an individual is threatened. Gartner predicts that by 2025, cyber attackers will use weaponized technology to harm or kill humans. Earlier, cyberattacks were restricted to things such as denial of services, information theft, and ransomware.

These cyberattacks have a heavy financial toll (billions of dollars), cause disruption in production, cause intellectual property to be stolen, and eventually, the brand reputation is tarnished. This is a never-ending battle in this digital age. Security vendors have come up with hundreds of security products and solutions to counter these cyberattacks. IBM has been at the forefront and is leading the security space with top-of-the-line products and solutions.

To understand the impact of a cyberattack, we just have to look a few years back at what happened with Ashley Madison. Ashley Madison was a dating app for those who were married, and the slogan they used to advertise then was *"Life is short. Have an affair."* Not surprisingly, the service had 37 million subscribers.

And then the unthinkable happened for the subscribers of the site. Ashley Madison used the weakest password encryption algorithm, and it was easily hacked. A hacker group called the Impact Group gave Ashley Madison 30 days to pay a ransom. As Ashley Madison did not pay, on the 30th day, they released about 60 GB of data with the names, email addresses, credit card numbers, and other details of the subscribers on the dark net. Soon, the media and the crooks started looking for famous personalities to hold them for ransom. The hack soon became public knowledge, leading to a large number of breakups, divorces, and even suicides. The financial implications of such breaches are unaccountable. The site and the brand of Ashley Madison were damaged permanently.

The point that needs to be understood from this scenario is that security breaches can cost lives, and hence any organization (whether it be a dating website, a bank, or a telecom company) needs to be on top of its game when it comes to security.

IBM QRadar is a solution suite that provides enhanced threat intelligence and insights into cyberattacks. These insights help organizations automate responses to threats and also help in devising new strategies to counter cyberattacks. An organization uses hundreds of enterprise solutions and security products from different vendors, such as firewalls, **Endpoint Detection Response (EDR)**, **Intrusion Prevention System (IPS)**, **Data Loss Prevention (DLP)**, and so on. IBM QRadar seamlessly integrates with all these products, consumes all the security data from them, and provides security alerts or insights that are actionable.

In this book, we will learn more about how to build your next-generation **Security Operations Center (SOC)** using the IBM QRadar solution suite. To understand **IBM QRadar** and how it functions, it is important to understand the different components. We call all these different QRadar components **managed hosts** (apart from the **Console**).

In this chapter, we will discuss various QRadar services for each component, which should be a good starting point to design the architecture for your SOC. As per different requirements, different components can be used in the deployment. Various aspects such as deployment types, scaling, upgrades, and licensing are discussed in corresponding chapters. In this chapter, however, we're going to cover the following main topics:

- Understanding the QRadar Console
- Exploring event data
- Exploring flow data
- Getting to know the Data Node
- Investigating QRadar components

Understanding the QRadar Console

The Console is the brain of QRadar and is the single indispensable component of QRadar. It can collect and process data and throw alerts based on the rules. This is the primary job of the Console. Other components (described later) are mostly used to scale these functionalities in one form or another. Now, let us look at the three major services running on the Console and understand them.

Tomcat

The primary utility of this service is for displaying the **User Interface (UI)** of QRadar. The QRadar UI can be accessed by typing the IP address or the hostname (if it can be resolved) in the browser.

If the `tomcat` service is down on QRadar, you will not be able to load the QRadar UI. It maintains the user sessions, active sessions, and current users – all those who have logged in to QRadar UI. It also plays a part in authenticating users, whether it is local authentication, **Lightweight Directory Access Protocol** (**LDAP**) authentication, or any other type of authentication. It is a multithreaded service that also deals with the export of data from the QRadar UI. The status of the `tomcat` service can be checked using a simple command:

```
systemctl status tomcat
```

We will cover troubleshooting tricks and tips for `tomcat` in the final chapter.

> **Important note**
> The Tomcat service is only available on the QRadar Console.

Hostcontext

When the `hostcontext` service is started, it triggers many other services with it. All the functionalities of QRadar are dependent on the `hostcontext` service. This service is part of all the QRadar managed hosts, unlike the `tomcat` service. The `hostcontext` service is responsible for replicating the deployment changes from the Console to other managed hosts.

The following is the list of services triggered because of `hostcontext`:

- `ecs-ec-ingress`: The specialty of this service from the following services is that even if the hostcontext service is stopped from the command line or if the `hostcontext` service crashes, `ecs-ec-ingress` keeps running and collecting events from Log Sources. If the `ecs-ec-ingress` service is stopped, there are two ways of starting:

  A. Restarting the `hostcontext` service

  B. Starting the `ecs-ec-ingress` service separately

- `ecs-ec`: The primary function of this service is to parse (map) the incoming events/flows. This service converts the events into a form that QRadar understands. The event is mapped to its event name by this service. For example, Linux OS has sent an authentication event to QRadar that there is an invalid user named `testdev` trying to log in via SSH.

  The payload of such an event would look like this:

  ```
  "Apr  10 18:26:40 servername sshd[26388]: input_userauth_
  request: invalid user testdev
  "
  ```

QRadar needs to make sense of this payload. This is called **parsing**. Then, QRadar needs to map this event to an appropriate event name, which is called **event mapping**. The event will be parsed as follows:

- Time: 6:26:40 p.m.

- Date: 10th April

- Event Name: `Authentication Failure`

- Server Details: servername

So, the two important functions of `ecs-ec` are as follows:

- Event parsing

- Event mapping

- `ecs-ep`: Once the events are parsed and mapped, they need to be processed. Rules are provided as a part of the initial installation. These rules can further be customized as per the security use cases of the organization. `ecs-ep` is responsible for matching each incoming event against all the enabled rules. If the rule conditions are fulfilled based on the incoming event/events, offenses (security alerts) can be triggered based on the rule action and rule response (defined).

  For example, we could have a rule to trigger an offense if we receive an event called `Authentication Failure` from Linux OS after 6 p.m. In such a case, looking at the previous event in the example, an offense will be generated.

  `ecs-ep` is also responsible for offense management in terms of the following:

  - Offense creation

  - Renaming offenses

  - Attaching events to triggered offenses

  - Offenses in a dormant condition

  - Offense closure

  - Offense deletion

- **QFlow**: This service is responsible for collecting flows in QRadar. Flows are network packet information collected in a specific format over a period of time.

- **Accumulator**: This service is responsible for creating global views in QRadar, which can be used for dashboards, reports, and so on.

- **Ariel proxy**: This service is responsible for relaying the search queries to the appropriate managed hosts.

- **Ariel query**: This service queries ariel databases across all managed hosts based on the query run on the Console.

There are many other services, and we will discuss them in detail when introducing the concepts related to them.

Hostservices

This service is also part of the Console and other managed hosts. Typically, QRadar has two types of databases:

- Ariel, which is not managed by `hostservices` (it stores event/flow data)
- Postgres, which is managed by `hostservices` (it stores configuration)

An ariel database has security data that is collected, and processed by `hostcontext`. Postgres has the configuration details of QRadar. This is managed by the `hostservices` service. Postgres is a **relational database management system** (**RDBMS**) that has multiple tables containing information on QRadar deployment, configuration, and settings. The Postgres database is replicated on different managed hosts using the `hostcontext` service. The Postgres database can be queried, if required, using the `psql` command line. We will discuss this in detail when we talk about optimization and tuning QRadar.

The brain behind QRadar is the Console, and the other components act as auxiliary parts of the system helping the Console perform the functions in a better way. Before jumping on to other QRadar components, let's first discuss and understand event data and flow data.

Exploring event data

Every organization that plans on building a SOC has hundreds and thousands of applications, servers, and endpoints that it would like to monitor. Each of these applications or servers has security and audit logs. These security and audit logs are what we call **event data**. QRadar supports multiple protocols such as Syslog, JDBC, and UDP multiline. It also supports product-specific protocols such as the Akamai Kona REST API protocol and the CISCO NSEL protocol. Using these protocols, QRadar can either pull data from Log Sources or we can configure Log Sources to send data to QRadar.

The data sent is in the form of events that are parsed and mapped to certain event names. The events can be viewed from the QRadar UI from the **Log Activity** tab. There are multiple query options that can be used.

The following figure shows a screenshot of the **Log Activity** tab with filters applied:

Figure 1.1 – Log Activity tab

Here, we refer to *Data Source* as *Log Source*, and they can be used interchangeably.

Event Processor

We previously learned that the Console is the brain of QRadar. However, there is a limit to the amount of data that can be collected and processed by the Console. This is where the **Event Processor** comes into the picture. The Console can delegate the work of collecting and processing event data to the Event Processor.

The main services running on the Event Processor are as follows:

- hostcontext: There are multiple services that are triggered when the hostcontext service is started. On each managed host (Event Processor), there is a configuration file named /opt/qradar/conf/nva.hostcontext.conf. It has a parameter called 'COMPONENT_PROCESSES'. Based on the values, the services are started when hostcontext is started. The main services that are part of the hostcontext service on the Event Processor are as follows:

  - ecs-ec-ingress

  - ecs-ec

  - ecs-ep

  - Ariel query server

- hostservices: This is like what we have seen on the Console.

You will see that there is no *ariel proxy service* on the Event Processor. This is because the ariel proxy service is only on the Console. When we search for data on the Console, the ariel proxy service sends the query request to the Event Processor. This request is accepted and worked on by the *ariel query server* service.

The ariel query server queries the ariel database, which is on the Event Processor, and sends the resultant data back to the Console, where it is shown in Log Activity.

> **Important note**
> We will be using the term *managed host* for any QRadar component that can be managed from the Console – for example, Event Processor, Flow Processor, Event Collector, and so on. There are very few QRadar components that are not managed by the Console. We will discuss them later.

An Event Processor can collect data using `ecs-ec-ingress`, parse the data using `ecs-ec`, and process the data using `ecs-ep`. You will notice that there is an ariel database in each Event Processor, which means that the *event data is stored locally*. Only when the data is searched is the resultant data sent to the Console for display. Over a period, the resultant data on the Console is removed as per configured policies

As the collection, parsing, and processing of data are done separately, whenever we add an Event Processor to the Console, we say that is a **distributed** environment.

Along with the Event Processor, the Event Collector also plays an important role in QRadar deployments. Let's discuss this further.

Event Collector

The Event Collector is the component that collects the event data and then either sends it to the Event Processor (if there is one) or to the Console for storage.

In some deployments, the Log Sources (which are configured to send data to QRadar) are in different time zones or different networks, and it is not feasible to send data directly to the Event Processor or Console. In such scenarios, the Event Collector can be used to collect local (to the network) event data, parse it using DSM, and send it to the Event Processor or Console for correlation and storage.

If the connection between the Event Collector and Event Processor is lost, event data is stored locally on the Event Collector, and when the connection is restored, the data is sent to the configured Event Processor.

Important services running on the Event Collector are as follows:

- `hostcontext`: These are the subservices of the `hostcontext` service:

  - `ecs-ec-ingress` – To collect event data

  - `ecs-ec` – To parse collected event data

- `hostservices`: Same as in the Console

We have discussed in detail the log data from the applications, servers, and endpoints. The other most striking feature of QRadar is its ability to process flow data. Let's discuss that in the next section.

Exploring flow data

Flows are different from events. Flow data is the information of the session between two hosts. For example, if an employee logs in at 9 a.m. and starts using social media, QRadar can capture these session details between the employee's machine and the social media site. This is done by capturing the network traffic from the span port of a switch. There are different types of flows, which we will discuss in detail later in *Chapter 4*. Flow data can be viewed in the **Network Activity** tab on the QRadar UI, as shown in the following screenshot:

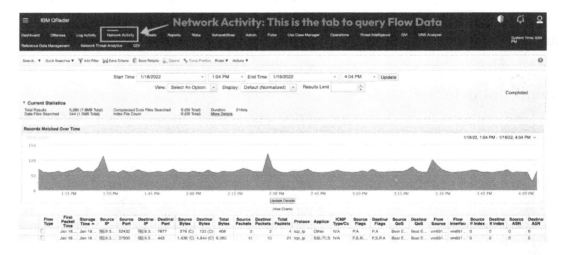

Figure 1.2 – Network Activity tab

Similar to an Event Processor, we have a Flow Processor and Flow Collector for flow data. Let's discuss them in detail next.

Flow Processor

Like the Event Processor, the Flow Processor is another managed host that collects and processes flow data. It has an ariel database where the flow data is stored, and it can be queried using the same mechanism as discussed for the Event Processor.

For the Event Processor, we had `ecs-ec-ingress`, which collected event data. But for the Flow Processor, we have the `qflow` service, which collects flows and then sends them to `ecs-ec` and `ecs-ep` for further processing and storage.

Important services running on the Flow Processor are as follows:

- `hostcontext`: For the Flow Processor, the `'COMPONENT_PROCESSES'` parameter in the `/opt/qradar/conf/nva_hostcontext.conf` file has different values than in the Event Processor.

- `qflow`: This service is responsible for collecting the flows. The Flow Processor does NOT have the `ecs-ec-ingress` service

- `hostservices`: Same as in the Console.

> **Important note**
> A single managed host can act as both an Event Processor and Flow Processor. For this to happen, you need to choose the correct option while installing. Usually, on enterprise-level deployments, Event Processors and Flow Processors are kept separate.

Flow Collector

Like an Event Collector, a Flow Collector is used to collect flow data, analyze it, and send it to the Flow Processor or Console for processing.

The Flow Collector has a special service like Flow Processor called `qflow`, which collects flows. Flow sources are defined on the QRadar Console UI and then the configuration is pushed to managed hosts, thus the Flow Collector understands which flows need to be collected.

Important services running on the Flow Collector are as follows:

- `hostcontext`: The subservices of the `hostcontext` service are as follows:

  - `qflow` – This service is responsible for collecting flows

  - `ecs-ec` – This service is responsible for aggregating and analyzing flows

- `hostservices`: Same as in the Console

Another important component usually used in huge deployments is the Data Node. Let's see why in the next section.

Getting to know the Data Node

Event and flow data are required for security purposes as well as for compliance. The amount of storage available on the Console and processors might not be enough for compliance.

For example, it may be mandated by Central Banks to keep event and flow data for 2 years. The available storage on processors can store data only for 6 months. In such a scenario, multiple Data Nodes can be added to a processor so that the processed data can be stored.

Adding a Data Node to deployment has two advantages:

- Increases the storage space for event and flow data
- Searches are more efficient when Data Nodes are used

Multiple Data Nodes can be attached to a single processor. One Data Node cannot be attached to multiple processors. What this means is that one Data Node will share data with just one processor.

When Data Nodes are added to the deployment, there is a process called **data rebalancing** that happens. The incoming data in the processor is distributed amongst the Data Nodes that are attached.

If a Data Node goes down (or crashes), the incoming data is not written to the Data Node. Once the Data Node is up, data is again rebalanced between the processor and Data Node. We will touch more on Data Nodes while discussing searches in *Chapter 6*.

Other QRadar components

Along with the collectors, processors, and Console, we have a few more feature-rich components that we will look at next. Some of the components that we will discuss are managed hosts, while the QRadar Packet Capture component is an unmanaged host.

Let's get started and discuss each component in detail.

QRadar Network Insights

QRadar Network Insights (**QNI**) is another QRadar component that is used for capturing flow data. Now, it might be interesting to know why we have QNI when we already had a Flow Collector and Flow Processor, which were also meant to capture flow data.

The Flow Collector and Flow Processor have a service called qflow, which is responsible for collecting flow data. qflow captures the first 64 bytes of the packet and not the complete payload. This is done to extract important information from packet data. If complete packet data is saved on the Flow Processor, the amount of network packet data would be so huge that it would fill up the Flow Processor in days, if not hours. So, it is practical to capture the first 64 bytes of packet information and store it.

QNI as an appliance (we will learn what an appliance is and how it is different from software installs in *Chapter 2*,) has special hardware to capture huge chunks of data, process that data, extract metadata, and send it to the processor for storage. So, if the same packet capture is sent to QNI, it can capture more data than qflow will capture with the first 64 bytes of information. For example, if the payload contains information that a user tried to access Facebook or tried to upload a file on Dropbox, QNI be able to capture the hash of the file that is uploaded (if available in the payload). This hash value can then be used by QRadar to determine whether it is a known threat by comparing the hash value of the file to the threat intelligence feeds it has.

In short, QNI provides us with more detailed information about the same packet capture compared to the information if Flow Collector/Processor is used using qflow.

In a single deployment, there can be multiple QNIs installed. QNI captures the packets and sends the information/insights to the processor/Console in IPFIX format. As the insights from QNI are more detailed, granular rules can be created, providing better security coverage.

QRadar Incident Forensics

Till now, we have talked about managed hosts, which are deployed in the QRadar environment for collecting and processing data in terms of events and flows. When it comes to forensics and deep diving into a security incident, **QRadar Incident Forensics (QRIF)** comes into the picture.

A security incident, such as financial fraud, would require an in-depth investigation of all activities that took place over a period of time. QRIF helps in reconstructing network sessions from the packet capture available.

For example, last Friday night, someone in a bank environment downloaded a script on the server and ran it. After a while, some of the accounts were compromised and a transfer of funds took place. From events and flows, a preliminary analysis can be done. But if there was complete packet capture available from the span port on the switch where the server was connected, QRIF could recover and reconstruct the complete session based on the source and destination IP address (or any other query), and a web session showing the exact actions of the threat actor can be seen.

QRadar Packet Capture

If QRIF can reconstruct the complete sessions, it will require complete packet captures to be stored from the time of the incident. This is where **QRadar Packet Capture** (**QPCAP**) comes into the picture. QPCAP is not a managed host. This can be placed in the network where the capture port of QPCAP can collect the network packets and store them. Network packets can fill up QPCAP very quickly. So, it has large storage space and special hardware for direct I/O disk access.

Typically, QPCAP is configured to capture and store network packet data for 7 days. If needed, multiple packet captures can be used as per the requirement.

QRadar Vulnerability Manager

Along with event data and flow data, another important aspect of information from the point of view of security is the vulnerability data of the assets in the environment. Administrators should have information on the assets available and how vulnerable these assets are. **QRadar Vulnerability Manager** (**QVM**) performs active and passive scans on the assets in the network.

For example, a QVM scan can be run periodically to understand whether the patching of all Microsoft servers is completed. It helps in determining the security posture of an organization.

QVM is a complete asset management tool right from discovering assets to running remediation reports for those assets. For this, QVM seamlessly integrates with other products such as BigFix and IBM SiteProtector.

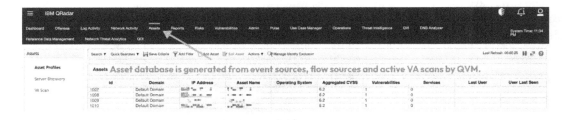

Figure 1.3 – Assets tab

In the preceding figure, the assets database can be seen. Each asset will have a unique ID, domain information, and vulnerability information associated with it.

Figure 1.4 – Vulnerabilities tab

In the preceding figure, a scan is run twice over a period. There are eight assets that were scanned both times, but the number of vulnerabilities is different. This indicates that over a period of time, new vulnerabilities were found for the same assets.

QRadar Risk Manager

QRadar Risk Manager (**QRM**) is a managed host that collects configuration information from the network devices in the environment. This information is then pictorially depicted in the form of network topology.

For example, in a typical data center, there are multiple network devices such as routers, firewalls, intrusion prevention devices, and so on. QRM pulls the configuration details from all the network devices and creates network topology graphs. It can also simulate any changes in the policies by calculating the risk to the environment.

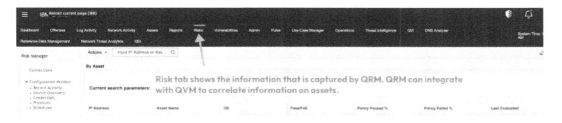

Figure 1.5 – Risks tab

QRM can integrate with QVM. Risk policies can be defined on QRM. To give an example of a risk policy in QRM, a policy can be defined to find out whenever there is web traffic, but it is not HTTPS traffic. That means, whenever there is web traffic on HTTP, the policy triggers, and the information related is collected by QRM. This information is useful to change the vulnerability risk scores of the assets involved. Whenever risk scores are changed in QVM, alerts can be generated via email or dashboard alerts to get higher visibility on the high-risk assets.

The **Risk Manager** tab shows information about the network devices in the environment. In the upcoming versions of QRadar, QRM and QVM will be part of a single appliance and will be sold as a single solution. But the fundamentals of both QRM and QVM remain the same, as discussed in this chapter.

QRadar App Host

On QRadar, just like on our Android or Apple phones, there are applications, or apps, that can be installed. These applications enhance the usability of QRadar and also provide additional features. All the QRadar apps are publicly available on the QRadar App Exchange portal. Here is the link to the portal: `https://exchange.xforce.ibmcloud.com/hub`.

In the *Understanding the Console* section, we covered so many services and tasks that are running on the Console. Installing multiple apps on the Console will add computational overhead in terms of CPU cycles, memory utilization, and the disk space required by the apps. For this reason, we have a component called App Host, where the Console offloads all the apps.

In the following figure, we can see how different components fit together. The QRadar Console is the most important component to which other components are connected for command and control (except QPCAP).

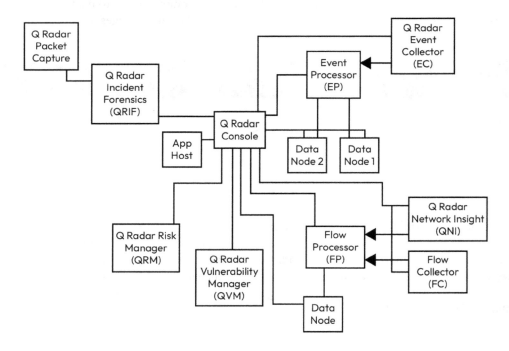

Figure 1.6 – Sample QRadar deployment showing multiple components

From *Chapter 8* onward, we will discuss at length all the important QRadar apps that can be installed on QRadar.

Summary

This first chapter was all about understanding the different IBM QRadar components. Being a highly scalable product, it is imperative that we know which components are to be used to scale the solution. We also understood how the event and flow data is handled in QRadar by different QRadar components. With that, you now have an idea of the basic architecture of QRadar components and how they fit together.

In the next chapter, we will learn how these components work together.

2

How QRadar Components Fit Together

After exploring the different QRadar components, let us now understand how they fit together. There are different ways to design a QRadar deployment, and it completely depends on the scope of the security operations that we intend to undertake. It depends on the number of endpoints, applications, users, and servers that we want to bring under the radar.

Therefore, in this chapter, we will explore two different types of deployments and why and when they are used. The following are the two types of QRadar deployments that we will discuss:

- All-in-one deployment
- Distributed deployment

We will also discuss the conditions under which the different components are added to the deployment.

All-in-one deployment

In this type of deployment, the processing of the data is done only by the Console. So, there will be no data processors (other than the Console) in this type of deployment, and there will be no separate Event Processor or Flow Processor in the deployment either. Processors, as we know, are responsible for processing and storing data. In **all-in-one deployment**, the aim is that the Console would be the only component where we will have the data stored or have a Data Node attached to the Console.

Event and Flow Collectors can be added to this type of deployment. Also, QNI can be added, which will collect flows and send data to the Console for processing.

The following diagrams are a few examples of all-in-one deployment.

In the following example, we see a QRadar Console that is collecting data from different Log Sources and Flow Sources. We also have QRadar UI access via a web browser. Administrators and analysts can log in via the QRadar GUI to configure data collection, monitor events, and work on security alerts, also known as **offenses**.

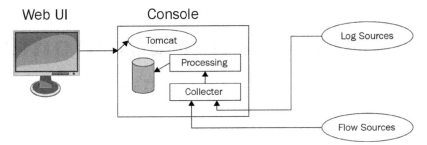

Figure 2.1 – All-in-one deployment 1

In the following diagram, we see the QRadar Console with an Event Collector and Flow Collector. Here, instead of the Console collecting the data directly, the Event Collector and Flow Collector are collecting the data.

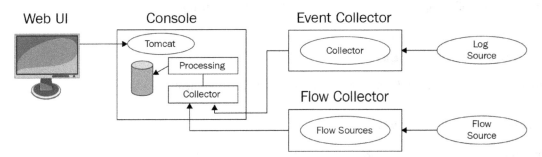

Figure 2.2 – All-in-one deployment 2

In the following example, we see that, along with the Log Sources and Flow Sources, we have another component, known as App Host. As discussed in *Chapter 1*, App Host does not store any event or flow data. Therefore, even after adding the App Host, this type of deployment will still be called *all-in-one*.

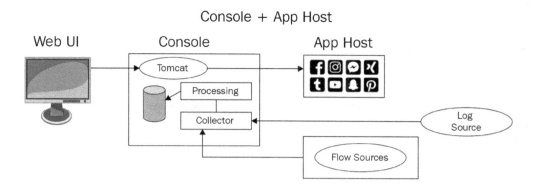

Figure 2.3 – All-in-one deployment 3

App Host is required when we need a considerable number of apps to be installed. Apps such as **User Behavior Analytics** (**UBA**) and Watson Advisor are computationally heavy applications using ML and AI models. They need significantly more CPU processing and memory. This may put a tremendous load on the Console, which is already performing so many activities.

Apps are nothing but Docker images running as containers in QRadar deployment. QRadar provides a framework for these apps to work independently of each other.

We will discuss apps in detail from *Chapter 8* onward.

Distributed deployment

In all-in-one deployment, the heavy lifting work in terms of data processing and storage was done by the Console. Once processors are added, we add more processing power and more storage. This helps the Console to free up resources for other important tasks.

In huge customer deployments where terabytes of data are processed daily, using all-in-one deployment will not suffice. We need more processors to correlate data and store it. Each processor comes with individual storage capacity. For example, for one of the biggest deployments of QRadar, which processes around 2 TB of data on a daily basis, we have 3 Event Processors. These Event Processors are in high availability, which means that for each primary Event Processor, we have a corresponding secondary Event Processor present. Each of these three processors shares the load of correlating the incoming events to that particular processor. So, on average, each event's average size could be around 500 bytes, which translates to around 2 TB divided by 500 bytes/event = 4,000,000,000 events per day. Moreover, 4,000,000,000 events per day equate to 46,296 **events per second** (**EPS**).

> **Important note**
> We have discussed Event Processors here, but the same calculation and understanding work for Flow Processors too.

For the Console to process around 45,000 EPS, it would involve very heavy lifting. It would be advisable that more processors be added to share the load of correlation and storage. It is recommended that, depending on the specifications of the hardware of the processor, we may add 2 or 3 processors to share the 45K EPS. If we select 3 processors, each processor can then proceed with 15K EPS.

> **Important note**
> The hardware specifications decide the amount of EPS license that should be applied to the processor.

So, a distributed deployment is when processors are added to the deployment. In large-scale enterprise solutions, the deployments are usually distributed. For example, Event and Flow Processors are sometimes geographically distributed. A Console could be in North America while the Event Processors could be in Australia, India, the Middle East, and Europe.

The following diagram shows a simple scenario of distributed deployment:

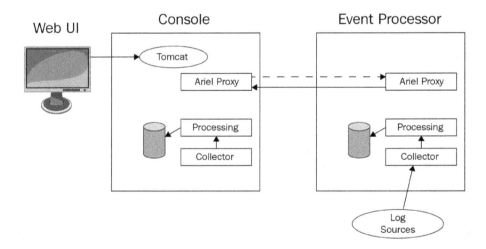

Figure 2.4 – Distributed deployment

In the preceding diagram, an Event Processor is added to the deployment. The event collection and processing are done on the Event Processor. When a search is run from the QRadar UI, that is, from the Console, the Tomcat service sends the query to the Ariel proxy service, which, in turn, sends the query to all the managed hosts (in parallel) that have the Ariel Query server running. On the managed host, in this case, we have an Event Processor where the query is sent. The Ariel Query server collects the log from the local database (here, the Event Processor's Ariel database) and sends the resultant data back to the Console, where it is displayed as the search result.

On a similar line, a Flow Processor can be added to the deployment.

Now, let us understand some intricacies of geographically-distributed deployments.

Let's take a step back to deep dive into event data. Till now, we understand that we collect event data from different Log Sources, and this data can be searched and displayed in the **Log Activity** tab of QRadar. If you double-click on any event, we can see event details:

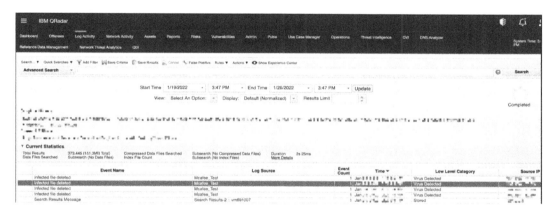

Figure 2.5 – Screenshot of Log Activity

In the preceding example, we see an event, with event name– **Infected file detected**. From the **Log Activity** tab, you can filter the number of columns that you would like to see. By default, for an event, we have **Event Name**, **Log Source** (to which the event is associated), **Event Count** (this deals with coalescing of events, which we will see in later chapters), **Time** (of the event), **Low Level Category** (of the event), **Source IP**, and **Source Port**.

One of the important factors that we will discuss is the time of the event. By default, the start time is shown in **Log Activity**:

- **Start Time** is the time the event hits the QRadar Event Pipeline. So, it is the time the ecs-ec-ingress service receives this event.

- **Storage Time** is the time when the ecs-ep process stores the event in the Ariel database.

- **Log Source Time** is the time the event is created in the Log Source.

Now, let us look at the event time details in this example:

Figure 2.6 – Event details and different time factors

From the preceding screenshot, we see that **Log Source Time** and **Start Time** are different. This means that the event was created way before it was sent to QRadar for parsing and processing. There may be different reasons for that. One of the reasons could be the protocol used to fetch logs had issues. For example, if you are fetching logs using the **Java Database Connectivity** (**JDBC**)protocol, there are marker files that determine which data needs to be pulled. If the marker file gets corrupted, it will need to be reset. After resetting, the JDBC protocol will start working. But this will create a time difference between **Log Source Time** and **Start Time**. The main takeaway from this example is that **Log Source Time** can be different from **Start Time**.

QRadar can convert the time from the payload to the time zone in which processing is done. So as per our example, the QRadar Console is in the India time zone and the event collection happened in the Japan time zone. When the Event Collector in Japan sends the event to the Console for storage, the Log Source time will be converted to the India time zone. If this event is part of an offense (alert) on QRadar, then it is very useful if the events are in a single time zone to understand the chronology of the events.

Another aspect while designing distributed deployments is the amount of event data and flow data collected and processed. A Console has limitations based on the EPS and **flows per minute** (**FPM**) it can process.

There are limitations to the EPS and FPM that can be processed by the Console. If an organization needs to collect and process more data than say 30K EPS, we will need to add a new Event Processor. Similarly, for flow data of more than 1.2M FPM, we will add a new Flow Processor to the deployment.

Another aspect to look at is the retention period set for the event and flow data. The retention period is configured as per the business need or if there are any regulations that need to be followed. For example, for a certain state, it is mandated that financial institutions such as banks should retain at least 2 years of events and flows. In such a scenario, even though the EPS is well within the capacity of the Console to process, there are space constraints on the Console. And for this space constraint, it is imperative to have Event and Flow Processors (as needed).

In the previous chapter, we have seen that there are other components in QRadar. Let us now understand how they fit together.

QRadar Incident Forensics (QRIF)

The QRadar Incident Forensics processor is added to the QRadar Console like any other managed host. Packet capture appliances are attached to the QRIF processor. This helps the processor import the required packet capture information from **packet capture (PCAP)** appliances.

Usually, when we are investigating an offense, we would like to see what happened, which in essence is incident forensics. For example, an offense is triggered for an employee called Bob. If we right-click on **Bob** and click on **Forensic recovery**, the QRIF processor will collect all information related to Bob, such as the emails that Bob sent, his downloads, chat sessions, browser history, and so on.

The QRIF processor queries the data based on the filter from the PCAP appliances and recreates the traffic as it is. So, the email attachments that Bob sent are also reconstructed.

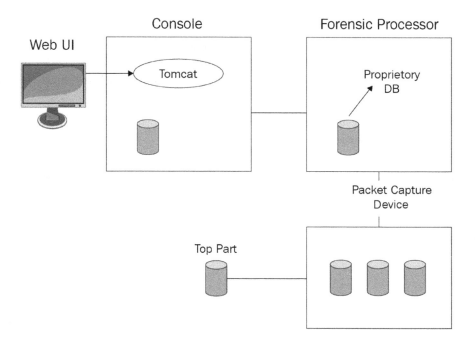

Figure 2.7 – Deployment with QRIF and PCAP

A QRadar deployment can have multiple QRIF processors and multiple PCAPs attached to it. The PCAP appliances capture the network traffic from the tap port on the switch, router, and so on.

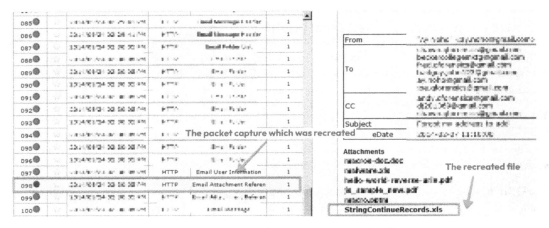

Figure 2.8 – Screenshot of QRIF capturing packet data

The preceding figure is a snapshot of the **Forensic** tab in the QRadar UI. Here, we have recreated a PCAP (number 98). From the PCAP, all the email attachments can also be retrieved. We can see an XLS file recovered, and its details can also be viewed. This helps in the forensic investigation during a security incident.

Like any other managed host, multiple QRIF processors can be added. And to each managed host, multiple PCAPs can be attached.

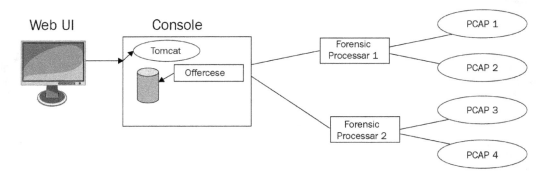

Figure 2.9 – Deployment with multiple QRIF processors and PCAPs

The preceding diagram shows multiple QRIF processors in a deployment.

> **Important note**
>
> In a QRadar deployment, the version of the QRadar Console and all other managed hosts is the same. For example, if the QRadar Console is on version 7.5, all the managed hosts should also be on version 7.5. QRadar PCAP is not a managed host and the QRadar Console does not manage it. Hence the version of the QRadar Console and QRadar PCAP could be different.

In this section, we have covered how QRIF is placed in a QRadar deployment and discussed in detail how it works. In the next section, we will see how QRM is placed in QRadar deployment.

QRadar Risk Manager

QRadar Risk Manager (**QRM**) is another managed host that can be added to QRadar deployment. For each deployment, there can only be one QRM. QRM uses network topology information such as configuration backup from network devices, and vulnerability data from events to understand and prioritize the risks.

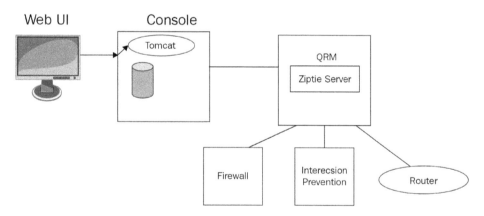

Figure 2.10 – Deployment with QRM

In *Figure 2.10*, we see that QRM is connected to network devices such as firewalls, intrusion prevention systems, and routers. QRM has multiple features that can make the life of QRadar admin easier. It provides a "discovery" feature. The admin will need to provide the IP address range. QRM will then start as a scan and will discover all the network devices in the subnet. Multiple IP address ranges can be provided, which would make the discovery of different network appliances possible.

> **Important note**
> QRM works on layer 3 of the **Open Systems Interconnection (OSI)** model.

Once discovered, the configuration details from the network appliances can be imported using a scheduled backup. The configuration details of the network appliances provide QRM with the required information to create network topology diagram.

QRM has different adapters, using which it can import the network configuration from network devices. For example, the adapter for the F5 firewall would be different than the adapter for the CISCO router. QRM can support multiple network appliances using their respective adapters.

Ziptie Server is used to collect all the configuration backups and normalize the data so that it can be used by QRadar. Based on the network configurations, the policies of different devices can be compared. If there are missing policies, alerts are sent in the form of offenses.

One of the great features of QRM is to simulate an attack on network assets. The result of the simulated attack can be used to understand which assets are at risk and how to remediate them. Another type of possible simulation is to change network policies on the network appliances and then simulate attacks.

Some organizations have specific compliance requirements, such as **Health Insurance Portability and Accountability Act (HIPAA)** and **General Data Protection Regulation (GDPR)**. QRM can help to automate the compliance checks too.

QRM is a managed host and, like all managed hosts, it has the same version as that of the QRadar Console.

Summary

In this chapter, we have discussed how all-in-one deployment is different than distributed deployment. We also looked at modern security challenges of integrating multiple event, flow, and asset sources together to understand the security posture of your organization. Using all the QRadar components discussed, we can build a QRadar environment suitable for our security needs. We have discussed different QRadar components at length with diagrams to understand how they integrate with each other. Hopefully, along with the previous chapter, you now have a clearer idea of what the overall architecture of QRadar may look like.

In the next chapter, we will dive deep into the different types of QRadar deployments. Also, we will discuss how to scale and upgrade QRadar. The discussion will also include the use of a license in QRadar.

3
Managing QRadar Deployments

Security threats are evolving every day. You will find threats in cloud solutions where your applications are installed, or threats present in the form of insider threats. With a zero-trust approach, we always assume that we are being attacked. Therefore, the security solutions we use should scale up, upgrade, and integrate new technology effortlessly. For emerging threats, new features, applications, and extensions are released in QRadar regularly. Sometimes, even security products are vulnerable. For this, we need to apply security patches on these products. These security patches are provided by IBM from time to time. The aim of a QRadar admin should be to keep the QRadar system up to date by upgrading to the latest available versions.

As new products are integrated, more data is integrated into QRadar, and for that, appropriate changes are required relating to the licenses installed. A few applications may need special licenses to work.

The following are the points that we will further discuss in this chapter:

- Understanding different types of QRadar deployments
- Installing QRadar
- Upgrading QRadar deployments
- Scaling QRadar deployments
- Licensing

Understanding different types of QRadar deployments

Till now, we have covered QRadar components and how they interact with each other. The different types of deployments depend on the platform on which QRadar components are deployed. In the following subsections, we will cover the five types of deployments.

QRadar appliances

IBM works with Lenovo and sells QRadar hardware in the form of M4, M5, and M6 boxes. These boxes have a preloaded ISO that already has RHEL and QRadar code. You do not need to install RHEL separately, which saves a lot of time. An ample amount of time can be required to install RHEL and configure partitions as per requirement. While installing QRadar, you can determine what component of QRadar you want to install on the box, whether it is a QRadar console or any managed host.

QRadar installed on virtual machines

QRadar can also be installed on virtual machines. Before installing, ensure that the hardware requirements are met – for example, the memory and CPU requirements. There are two types of installations possible on virtual machines:

- RedHat can be installed separately and then partitioned as per requirement. Then, install QRadar on top of RHEL.
- Install using QRadar ISO, which has embedded RHEL code. This is a much simpler approach, as partitioning is taken care of. The following link provides detailed steps on the virtual appliance installation: `https://www.ibm.com/docs/en/qsip/7.5?topic=installations-virtual-appliance`

QRadar on bare-metal servers

If you have bare-metal servers, ensure that the hardware requirements are met, and then install QRadar using the QRadar ISO; alternatively, install RHEL and then QRadar on top of it.

For this, you will require access to an appliance that uses KVM technology, such as an **Integrated Management Module** (**IMM**). Using this, you can mount the QRadar ISO and begin the installation. We would recommend that you use the QRadar ISO rather than installing the RHEL OS and then install QRadar on it.

QRadar installation on cloud solutions

If you have all the applications and data on a cloud solution, you have the choice to install QRadar on your cloud environment. QRadar is supported on the following:

- An Amazon Web Services cloud solution
- Google Cloud Platform
- IBM Cloud
- Microsoft Azure

QRadar has marketplace images available that can be installed readily on your cloud solution. You can also install managed hosts on these cloud solutions. This makes it convenient to manage data and security within the cloud.

QRadar Community Edition

IBM provides a free version of QRadar for everyone to download and install on a virtual machine, bare-metal server, and so on. Students and security professionals alike can install it to understand how QRadar works and gain first-hand experience with it. It is a fully featured version, though the **events per second** (**EPS**) and **flows per minute** (**FPM**) capacity for this version is 50 EPS and 5,000 FPM. As the throughput is low, the resources required are also low and can even be installed on a laptop that has specified resources.

You can download QRadar Community Edition here: `https://www.ibm.com/community/qradar/ce/`

In the previous section, we understood the different ways in which QRadar is installed. Now, let us look at a sample installation in detail.

Installing QRadar

The first step of QRadar installation is to understand which QRadar components will be needed in the deployment. We can always add processors and collectors when needed. However, we must start with the first step, which is requirement gathering. Let's get started:

1. Understand the amount of data in terms of EPS and FPM that will be ingested in QRadar. This will help us understand the number of processors required.

2. Check whether the data has to be collected from geographically different regions, and also check the data privacy rules of the countries/states. This helps us to understand whether we will need processors or collectors. If the bandwidth is pretty low for the remote sites, we may also need **Disconnected Log Collector** (**DLC**).

3. As per the EPS and FPM, ensure that hardware and software requirements are met. Ensure that the machine has enough memory, CPU, and storage before starting the installation.

4. Understand the data retention policy of the organization, and accordingly, you will be able to calculate the amount of storage required.

5. Check whether you may require QRadar Risk Manager, QRadar Vulnerability Manager, QRadar Incident Forensics, and so on. Check the hardware requirements for each appliance.

Important note

Here are the pre-requisites for installing on your hardware `https://www.ibm.com/docs/en/qsip/7.5?topic=installations-prerequisites-installing-qradar-your-hardware`

Here are the system requirements for virtual appliances: `https://www.ibm.com/docs/en/qsip/7.5?topic=installations-system-requirements-virtual-appliances`

Once we are good with the software and hardware requirements, the next step is to download the QRadar ISO from Fix Central (this can be done at `https://www.ibm.com/support/fixcentral/`). You will need to log in using your IBM ID and download the ISO.

In the previous section, we discussed different types of platforms and different ways to install QRadar. The following are the steps you need to take when you install on a virtual machine. Note that these steps are pretty similar to when you install QRadar on a bare-metal server. Let's get started:

1. Next, you'll need to select the type of setup. For a first-time installation, you should select **Normal Setup**. **HA Recovery Setup** is used when an HA appliance needs to be reinstalled.

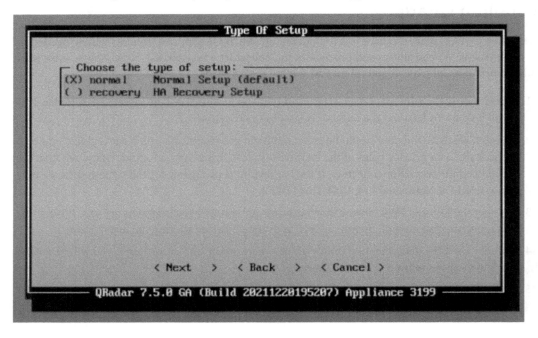

Figure 3.1 – Selecting the type of setup

2. Then, insert the date and time, and you can even add the server details of the NTP server.

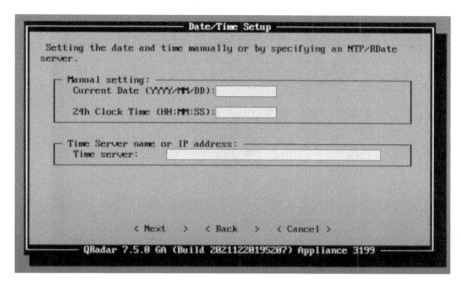

Figure 3.2 – Setting the date and time

3. Select the time zone.

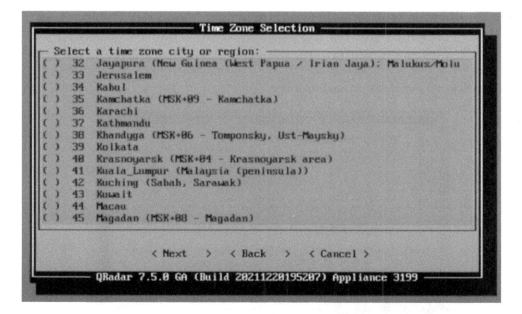

Figure 3.3 – Selecting the time zone

4.   Select how you would like to connect the QRadar appliance. These are the network settings, which also include an option to bond interfaces.

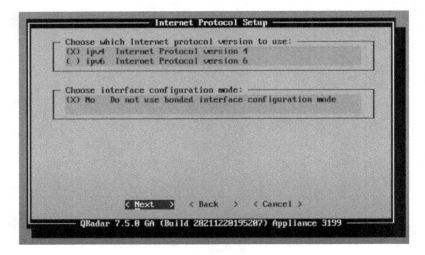

Figure 3.4 – The network settings

Important note

More about bonding interfaces can be found here: `https://www.ibm.com/docs/en/qsip/7.5?topic=installations-configuring-bonded-management-interfaces`.

5.   Then, select the available network interface.

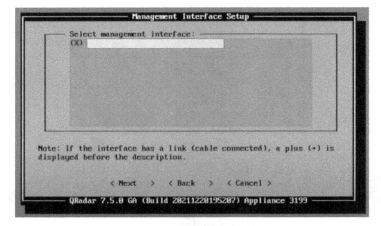

Figure 3.5 – Selecting the available network

6. Then, insert the hostname (i.e., the **Fully Qualified Domain Name** (**FQDN**)), IP address, gateway address, DNS address, and so on.

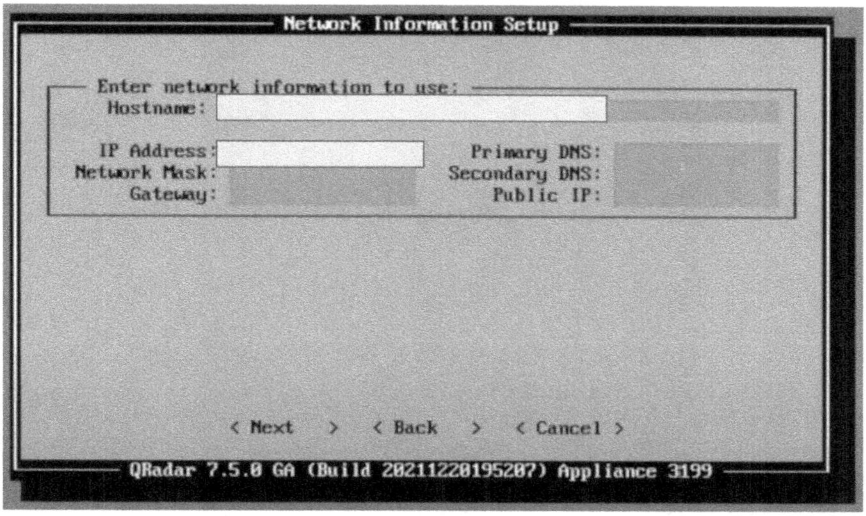

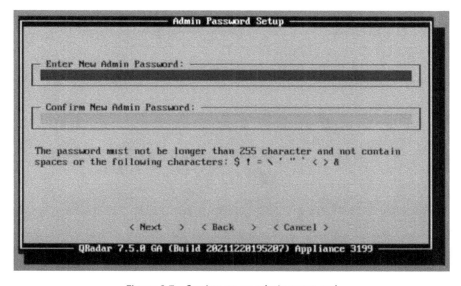

Figure 3.6 – Adding network information

7. Then, enter the admin password for the QRadar deployment (if it is a console). The admin password is used for QRadar **User Interface** (**UI**) access via the browser.

Figure 3.7 – Setting up an admin password

8. Then, enter the root password, which will be required when you log in to the QRadar CLI.

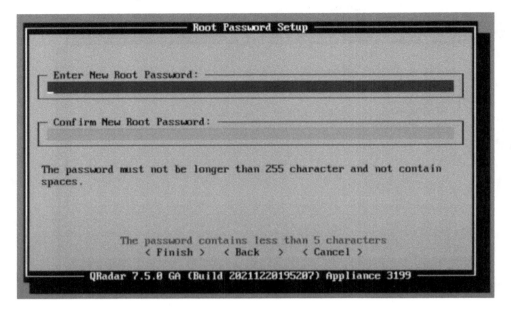

Figure 3.8 – Setting up a root password

9. The installation will continue, and once completed, you will get a prompt that the initial configuration and installation are completed.

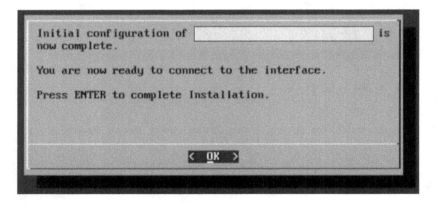

Figure 3.9 – The installation completed prompt

In brief, these are all the steps involved in the installation of QRadar components. Now that you are familiar with the different types of deployments and how to install them, let's discuss the two interesting and must-know aspects of QRadar – upgrading and scaling. Rest assured that if you upgrade QRadar as per the instructions, it is always a smooth process.

Upgrading QRadar deployments

IBM QRadar releases major version upgrades once a year and minor upgrades every quarter. The minor upgrades are also called Fix Packs or Upgrade Packs, depending on which version of QRadar you are upgrading. Major version upgrades usually arise with changes to operating system kernels, and QRadar may require a reboot after or during the upgrade. If minor upgrades also contain OS updates, a restart may also be required for minor upgrades. If QRadar Support encounters any defects and they need to be backported to an older version, **Intermediate Fixes** (**IFs**) are usually released. These IFs are version-dependent. So before installing any updates, always go through the release notes for that version.

Many times, the estimated load on QRadar in terms of events and flows processed may increase. This happens when more devices are boarded in the customer environment – for example, multiple firewalls installed, many web servers added, and networking devices such as routers added. To accommodate the change in load, event processors and flow processors can be added.

Let us now dig deep into how to upgrade your existing deployment.

Upgrading the QRadar version

The first thing is to understand the current version of your deployment and the latest available version. It is recommended that we have the latest version or (latest – 1) version. QRadar release has many security updates, feature updates, OS updates, updates for the stability of the environment, updates to integrate new devices, and so on. This is why it is recommended to have the latest version to fully utilize QRadar's potential.

Once you have decided on which version to upgrade to, the next thing is to check the release notes of the QRadar version. Here, you will understand what new features, security measures, and updates are provided in this version. The release note also has comprehensive upgrade steps to carry out. It is highly recommended that you have **User Acceptance Testing** (**UAT**) environments where you can test out the upgrade. In huge environments where there are multiple managed hosts, UAT can be designed in such a way that it covers most of the upgrade scenarios. For example, if you have 20 event processors, 20 flow processors, 8 QNI appliances, and an App Host, then your UAT can have two event processors, two flow processors, one QNI appliance, and an App Host. This can simulate most of the testing that needs to be done. When the UAT deployment is upgraded, any shortcomings can then be noted. A support case can be opened with QRadar Support if there are upgrade issues or queries. Once the issues on the UAT are resolved, the best practices can be applied while upgrading the production deployment.

A few upgrades would warrant some extra steps to be taken. For example, to upgrade from QRadar 7.3.x to 7.4.2 and above, a filesystem migration is needed. Separate steps for the filesystem migration from GlusterFS to a **Distributed Replicated Block Device** (**DRBD**) filesystem are required. These are unique scenarios and may not usually be the case. The steps for the upgrade are straightforward and are described in detail in the release notes.

A comprehensive checklist for QRadar administrators can be found here: `https://www.ibm.com/support/pages/qradar-software-update-checklist-administrators`

Administrators should follow the guidelines for a pre-upgrade check using the preceding link and release notes for the upgrade steps.

It is always advisable that configuration backups are done before the upgrade. If there are major failures, a rebuild of the system may be required. When the system is rebuilt, a backup restore will be done. As long as the pre-upgrade checks are done and the upgrade instructions are followed, the upgrade will be successful.

There are different strategies for QRadar upgrades, including a **Patch All** option. I would not recommend this strategy for deployments where the number of managed hosts is more than four. When you select **Patch All**, the console is upgraded, and then the managed hosts are upgraded one after another. No manual intervention is required.

Another way to upgrade would be to upgrade the QRadar console first and then copy the SFS or ISO files to the managed hosts and upgrade them in batches. This strategy can be used for large deployments – for example, once a console upgrade is done, the next step would be to upgrade all the processors and data nodes in one go. The next step would be to upgrade event and flow collectors, including QNI, in one go.

In the second strategy, you can even move the event and flow collection to one set of event or flow collectors while other event and flow collectors are upgraded. For example, if you have 10 event collectors. You can divide the upgrade in two batches. First move the event collection to the remaining event collectors, say 5 of them. Upgrade the 5 event collectors. Move back event collection to the upgraded event collectors and then upgrade the remaining 5 event collectors. Once all event collectors are upgraded, event collection can be re-distributed as it was before upgrade. This ensures minimal downtime for collecting data.

Another point to note is that the upgrade files are usually SFS files and are hosted on the IBM Fix Central site (`https://www.ibm.com/support/fixcentral/`). They are categorized by version.

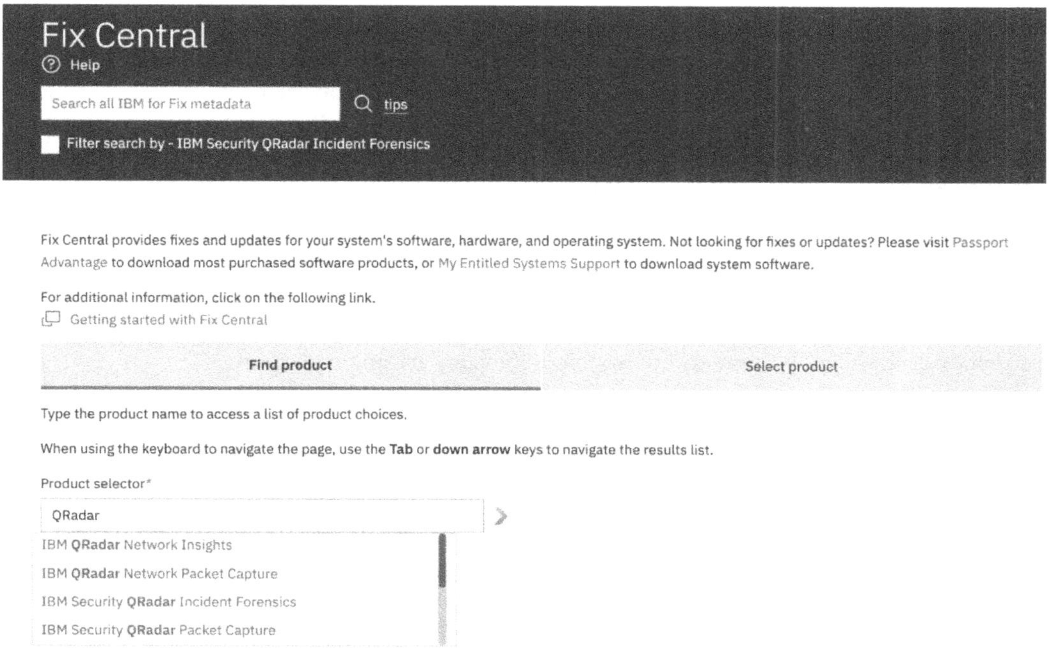

Figure 3.10 – The Fix Central web page

The SFS files for some QRadar components are different. The SFS files for the QRadar Console, processors, collectors, App Host, data nodes, QRM, QVM, and QNI are the same. For other products, such as QRadar Incident Forensics, the SFS files are different. All these files will be present on Fix Central.

Upgrading QRadar appliance firmware

In the earlier section, we discussed how to upgrade QRadar software. In this section, we will discuss firmware upgrades for QRadar appliances.

If you are working on QRadar, it may be either an IBM QRadar appliance or a bare-metal server on which QRadar is installed. For all the other types of deployment, apart from the aforementioned, firmware upgrades are not applicable. If you have not procured hardware from IBM, you should contact your OEM to keep your firmware updated. However, if you are using IBM QRadar appliances, customized firmware upgrades are available for the procured appliances.

IBM QRadar appliances can be categorized as M3, M4, M5, and M6 appliances, depending on the hardware capabilities provided. Each of these appliance types has a different version of firmware available. So, depending on the appliance type, you should upgrade to the latest firmware version available. The firmware files are either available as ISO files or IMG files. ISO files can be used to remotely upgrade the firmware using an IMM. IMG files can be used to upgrade firmware for on-prem appliances. A remote upgrade can save time if there are multiple appliances, as the upgrades can be started in parallel on different IMM consoles.

The details of how to perform a firmware upgrade are provided here: `https://www.ibm.com/support/pages/qradar-firmware-list-xseries-appliances`.

We also have an excellent blog on how to troubleshoot firmware upgrade issues:

`https://community.ibm.com/community/user/security/blogs/ashish-kothekar/2022/02/10/troubleshooting-qradar-firmware-upgrades`.

Now that we're familiar with upgrading, let's take an in-depth look at scaling next.

Scaling QRadar deployments

Every component has limited resources and, hence, a threshold capacity to collect and process data. Data, as we know, is ingested in the form of events and flows. So, every component has a limit on the number of flows and events that the component can collect and process. These thresholds are measured in terms of EPS and FPM. However, there are ways to scale your deployment. Let's see what they are in the following sections.

Scaling by adding data nodes

If data – that is, the events and flows – have to be retained for more time than planned, what is needed is more disk space. You can also add processors to the deployment and move a few log sources to the new processor. However, this is not the ideal solution in this case. What you should use is the data node. Adding a data node to a processor and/or the Console will not only provide more disk space to store data but also help searches to become faster.

Scaling by adding processors

In huge deployments, log sources are configured in such a way that data collection and parsing are done on event processors. The Console does other heavy-lifting processes such as reporting generations and handling searches. In such scenarios, if more data needs to be ingested, an event processor should be added. If the total incoming data is more than the license threshold, then a new license can be added to the license pool. We will see how licenses work in the final section of this chapter.

Scaling by adding collectors

In some scenarios, data must be collected from geographically diverse locations. The best solution, in this case, is to install event/flow collectors where the log sources are and then send the data to the processor or console to correlate rules and storage. Installing collectors at source helps by having a single connection from different geographies to processors. Otherwise, for each log source, depending on the type, different ports need to be opened across geographical firewalls.

When collectors are added, processors have more resources to process the collected data. Adding event and flow collectors helps QRadar perform better.

Scaling by adding CPU and memory on QRadar appliances

For the QRadar appliance, especially those deployed on virtual machines, it is easier to increase the computational resources by adding more CPUs and increasing the memory.

The CPU and memory should at least be what is suggested in the QRadar hardware prerequisites.

Till now, we have covered different types of deployments in QRadar and how QRadar can be upgraded as well as scaled. In the next section, we will discuss the finer points related to licensing in QRadar.

Licensing

Before the 7.3 version of QRadar, there were two types of licenses – one for appliance activation and one for the capacity of the appliance in terms of EPS or FPM. For QRadar version 7.4 and above, the only type of license that exists is for the capacity. All the licenses are added to the console, and it creates a license pool. From the license pool, a QRadar administrator can distribute the license capacity to processors.

Note that licenses are applied to processors and not to collectors. This is because even though data is collected by a collector, it is processed and stored by a processor. If you have a license of 30,000 EPS and the number of EPS received on the collector is more than 30,000, events can be buffered, parsed, and then sent to the processor where they are stored. There is a limit to the number of events that can be buffered. If the number of EPS is consistently more than the license threshold, the administrator will see a **License Threshold** warning on the QRadar Console. We will cover performance issues in more detail in *Chapter 11*.

IBM has a team that can help QRadar admins to source new or extra license capacity. They can send an email to q1pd@us.ibm.com, and the team will provide the required licenses. These licenses are paid licenses. QRadar admins can also contact their IBM Account representatives if licenses are required for **Proof of Concept** (**POC**) or other demos.

> **Important note**
> For QRadar Community Edition, which we discussed earlier, we do not require any licenses.

Another point to note while dealing with licenses is understanding the hardware of the processor on which it will be applied. For event collection and processing, there are two types of limitations that are applied – the first is the license limit and the other is the hardware limit. Every type of hardware has an upper threshold, beyond which data collection and processing have performance issues. So, a processor with a hardware limitation of 20,000 EPS should not be supplied with a 30,000 EPS license. This is because the processor will not be able to handle 30,000 EPS because of hardware limitations.

Summary

In this chapter, we have discussed in a very simplified manner how QRadar can be managed. It is important to understand the fundamentals behind how QRadar is designed to be upgraded and scaled. During an upgrade, do follow the pre-upgrade checks. Many times, the pre-upgrade checks can fail. In such scenarios, open a case with IBM Support. Do not upgrade if the pre-upgrade checks fail. For scaling of QRadar, there are two factors to keep in mind – licensing and which components to use.

Once you master the skill of managing QRadar deployments, you have eliminated more than 80% of the issues that could eat up your time and efforts as a team. You will have more resources available to focus on integrating your security solutions with QRadar. In the upcoming chapter, we will learn how to integrate different data in QRadar.

Part 2: QRadar Features and Deployment

Once the QRadar environment is built, it is time to understand the features it offers per your business requirements. Attaining this understanding is imperative before implementing the features. Ingesting data (events and flows), making sense of data (DSM editor), analyzing data, and then correlating it are all fundamental features of QRadar.

This part has the following chapters:

- *Chapter 4, Integrating Logs and Flows in QRadar*
- *Chapter 5, Leaving No Data Behind*
- *Chapter 6, QRadar Searches*
- *Chapter 7, QRadar Rules and Offenses*

4
Integrating Logs and Flows in QRadar

When an application is developed, a provision to log the details in it is also developed alongside. Logging is usually used to debug the application while developing as well as to troubleshoot and provide support to it. Every application can have different types of logs. Some of these logs contain security information, such as identity and access management logs, buffer overflow messages, and file tampering. All such logs play an important role in understanding the security risk for an organization.

Consider a scenario where a hacker gains access to a system; the first thing the hacker does is delete or purge the entries in the logs that would alert their unauthorized access to the system. This way, the hacker remains safe and can then do lateral movement to access other critical servers. So, how do we monitor and detect such attacks? The answer is by using flows in conjunction with logs. As the adage goes, *flows don't lie!*

If someone attempts to gain unauthorized access, flows instantly capture the packet data and send it for analysis to QRadar. This way, even if audit, access, or application logs are deleted on the end device by the hacker or anyone in general, the flows have already captured the activity. If we use **QRadar Network Insights (QNI)** to capture flows, then even layer 7 details (layer 7 or the application layer of the OSI model) are available for correlation. Event data and flow data form the basic data sources for QRadar correlation. Therefore, in this chapter, we'll be covering the following topics:

- Protocols and **Device Support Modules (DSMs)**
- Flows and types of flows
- **Disconnected Log Collection (DLC)**

Exploring protocols and DSMs

There are two factors we need to consider when integrating event data in QRadar:

- How to transfer data from end devices to QRadar

- How to parse or make sense of data that is received

Let's look at each of these considerations in detail.

How to transfer data from applications to QRadar

Different applications can be installed and run on different platforms or operating systems. For example, while some applications could be on a Windows machine on a bare-metal server, other applications could be running on the AWS cloud or some security appliance, such as a Ciscofirewall deployed in your organization.

All these endpoints are known as **log sources**, as we receive log data from them. As the log sources are on different platforms, they might also use different technologies and ways to log the data. For example, the AWS log sources can use something such as a S3 bucket, while a Linux security log would be saved on the same Linux machine by configuring a `/etc/rsyslog.conf` or `syslog.conf` file.

So, to integrate different log sources, different protocols are required. Broadly speaking, there are two types of protocols used in QRadar.

Active protocols

For these types of protocols, QRadar initiates communication with the log source. For example, an API request is made by QRadar to the end log source, and then the end log source sends the logs as per the request.

Another example could be a JDBC request initiated by QRadar for a database-related log source. The request is received by the database, and the query result is sent back as the response to the JDBC request.

Passive protocols

For these types of protocols, QRadar opens the required ports and waits for the log source to initiate the communication and data transfer. Typically, these are the syslog log sources, the streamer protocol from Cisco, and so on.

Some of the log sources, such as Cisco, have proprietary protocols that are used to fetch the event data. For cloud applications, there are protocols for a few applications, such as the following:

- The Microsoft Azure Events Hub protocol

- The IBM Cloud Object Storage protocol

- The Amazon Web Services protocol

As well as these, there is a Universal Cloud REST API protocol. This is used to fetch data from cloud applications using a REST API, where there are no protocols defined. The Universal Cloud REST API is defined using workflows, workflow parameters, connection details, and state. For any new application running in the cloud that has REST API support, this protocol can be used to modify workflows, parameters, and so on.

Once QRadar receives data, it is DSMs that come into the picture.

How to parse or make sense of data that is received

Once we have received data using any of the aforementioned protocols, the next step is to parse the data and make sense of it. The log sources collect data in different formats. QRadar needs to understand the data and extract the key values from the logs. This piece of code that parses the event data is called a DSM. There are hundreds of DSMs that are shipped with QRadar and supported out of the box.

For example, imagine that you have a CISCO AMP log source sending logs. QRadar will use the CISCO AMP DSM module to parse the events and make sense of the incoming data. If the events have an event name equal to Firewall Deny, the event/payload will be parsed and the normalized/parsed event will also be named Firewall Deny and have other information, such as the source and destination IP address, source and destination port, and time of the event. Therefore, DSMs parse or normalize the event payload so that it can be understood by QRadar and then used further to evaluate against the rules in QRadar. To explain normalization or parsing in detail, imagine that a payload has lots of information about the time of an event, username, error message details, error code, and so on, but QRadar cannot read the payload data as it is. Therefore, an attribute value pair is created by QRadar. From the payload, a username can be fetched and set as the attribute name. Similarly, other attributes can be fetched.

DSMs and protocols are thus critical components that are responsible for the collection and parsing of event data. DSMs and protocols are pieces of code, and this is updated when a new version of end log sources is upgraded.

For example, QRadar has a DSM for Oracle's Audit Vault product. Oracle's Audit Vault has been upgraded to the next version. The IBM QRadar development team releases the new version of the DSM for Oracle's Audit Vault product. There is a feature in QRadar called AutoUpdate, which can help download the latest version of DSMs and install them on the QRadar system. This AutoUpdate feature can be customized to download the latest version but wait for human intervention for installation, and so forth. Similarly, with QRadar upgrades, whether it is a minor or major upgrade, new versions of DSMs as well as protocols are downloaded.

> **Important note**
>
> When needed, even DSMs and protocols can be upgraded manually. There is a centralized repository for QRadar files called Fix Central. You can use the yum command to upgrade, install, or reinstall the package files. For details on how to use the yum command to manually work with the DSMs, you can refer to https://www.ibm.com/support/pages/qradar-using-yum-manually-install-reinstall-or-search-rpm-packages.

Every month, IBM QRadar releases a newer version of the DSM guide. The DSM guide is the documentation that lists all the products that can be integrated with QRadar for event logs. It has detailed information on how different protocols should be configured. Also, it has details for log source configuration. For example, a Linux log source will have information on how to use the syslog protocol to send RHEL logs to QRadar, while a VMware vCloud Director log source integration will need a REST API protocol to be configured.

At the end of the DSM guide, there is a comprehensive table that lists the products, their versions, the types of logs, and the protocols supported to integrate event logs in QRadar. You can follow the guide and configure the products as needed.

> **Important note**
>
> The settings mentioned in the DSM guide are tested thoroughly. Any change or modification in the log source configuration may affect the collection as well as the parsing of the logs. For example, If the Linux logs from RHEL need to be sent as syslog, you should not configure a REST API to collect Linux logs. It may or may not work, but it is definitely not supported by QRadar.

Services involved in the integration of an event log

In the previous section, we learned how to ingest event logs. The journey of an event log from being pushed to or pulled by QRadar to it being parsed, and then correlated and stored is called the **event pipeline**. An event pipeline can be imagined as a constant array of upcoming events in QRadar.

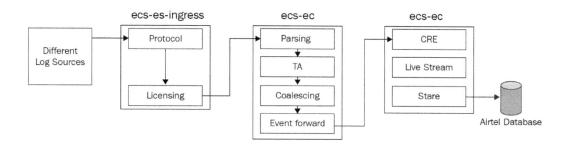

Figure 4.1 – An event pipeline with the service segregation

The preceding diagram shows how the events from different log sources are integrated into QRadar using different QRadar services.

The different shapes on the left-hand side of *Figure 4.1* suggest the different log sources, from where either the events are pushed by the log sources or pulled by QRadar. This depends on whether we use an active protocol or passive protocol. Protocols come under the `ecs-ec-ingress` service.

As we learned in *Chapter 3*, license that is required to collect and process the events. This license is the **Events per Second** (**EPS**) license. What this means is that QRadar can ingest only events for which it is licensed. Imagine that a company named Q has Qradar, which is licensed for 20,000 EPS. Now, this company has acquired a new company named R, which has about 100 new web servers as its most important asset. For the QRadar deployment, we will need to ingest logs from these 100 new Web Servers. Naturally, the number of EPS will increase.

Suppose the EPS grows to over 20,000 and remains constant over a few days or weeks. QRadar is designed to work with sudden bursts of events over a short period of time. However, for longer durations, we need to ask QRadar to process more than what it is licensed for. So, if the EPS is more than the licensed value over a significant amount of time, QRadar starts to drop the excess events and throws in a notification for the QRadar admin, stating that the incoming EPS is more that the licensed value. This is where the `ecs-ec-ingress` service jumps in with its licensing feature.

Once the event passes through the `ecs-ec-ingress` phase, the event is picked up by the `ecs-ec` service. The first operation that happens in the event is the DSMs kicking in. Using the DSMs, the events are parsed into consumable data. Let's look at an example:

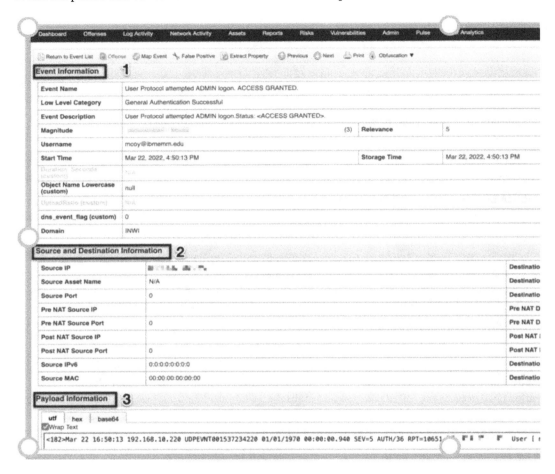

Figure 4.2 – Event details

In the preceding screenshot, we can see **Payload Information** at the bottom. Payload is the event that is sent by the end log source. Let's see how QRadar parses it.

QRadar understands that this event is for the access being granted. Then, digging deep, it also finds out the username under which this event has been logged. The username in this case is mcoy, and the user's email ID is mentioned as the username. Then, QRadar extracts the time from the event log. Finally, QRadar assigns a low-level category to this event so that we can make some sense of the payload. Everything that we have discussed is known as parsing the event.

So far, we have seen that QRadar supports different types of log sources. Imagine the amount of time required by a QRadar admin to integrate each and every log source manually. The QRadar admin will have to enter the end log source details, the log source identifier, the protocols used, and so on. To save time, QRadar came up with a feature called Traffic Analysis. If an event reaches `ecs-ec` and there is no log source that matches it, Traffic Analysis kicks in. It automatically creates a log source for such event logs under some specific conditions. More about Traffic Analysis can be found at `https://www.ibm.com/support/pages/qradar-understanding-traffic-analysis-and-log-source-auto-detection`.

Imagine a scenario where we have added a Cisco firewall as our log source and sent all the firewall events to QRadar. Now, for a firewall, there would be hundreds of events that accept the connection to pass through the firewall. The logs are kept usually from a compliance point of view, but this may or may not add value in terms of security when these events are parsed at QRadar. Hundreds of events would take up a lot of disk space, as they are stored in an Ariel database. QRadar has a feature to coalesce events, where if there have been more than 3 events in the last 10 seconds with a few similar event properties, such as source IP or destination IP, they are coalesced. What do we mean by coalescing? It means that if similar events come in, only one event is saved in the Ariel database – that is, a payload of the first event is saved for the rest of the events but payload details are not saved. QRadar keeps a track of similar events. This saves a lot of disk space.

> **Important note**
> The coalesced events are counted by the licensing module. So, even though a lot of events are coalesced, they are still counted when calculating the EPS.

The option to coalesce data is available on the log source management page for the log source. This means that you can selectively coalesce events from certain log sources. You can refer to this excellent article on how QRadar coalescing works: `https://www.ibm.com/support/pages/qradar-how-does-coalescing-work-qradar`.

Another feature of QRadar is event forwarding. There are many options when we configure events to be forwarded to another deployment or another event collector or processor. The other options are to drop the events or bypass **CRE (custom rule engine)**correlation. Let us consider the same example of the Ciscofirewall to explain one of the options in the event forwarding feature. Suppose QRadar receives too many events with different event names from the Ciscofirewall. As the event names are different, the events will not be coalesced. However, storing these events will eat up a lot of disk space. In such a scenario, you can set up an event filter and choose to drop such events. The screenshot in *Figure 4.3* shows the options available when we perform event forwarding.

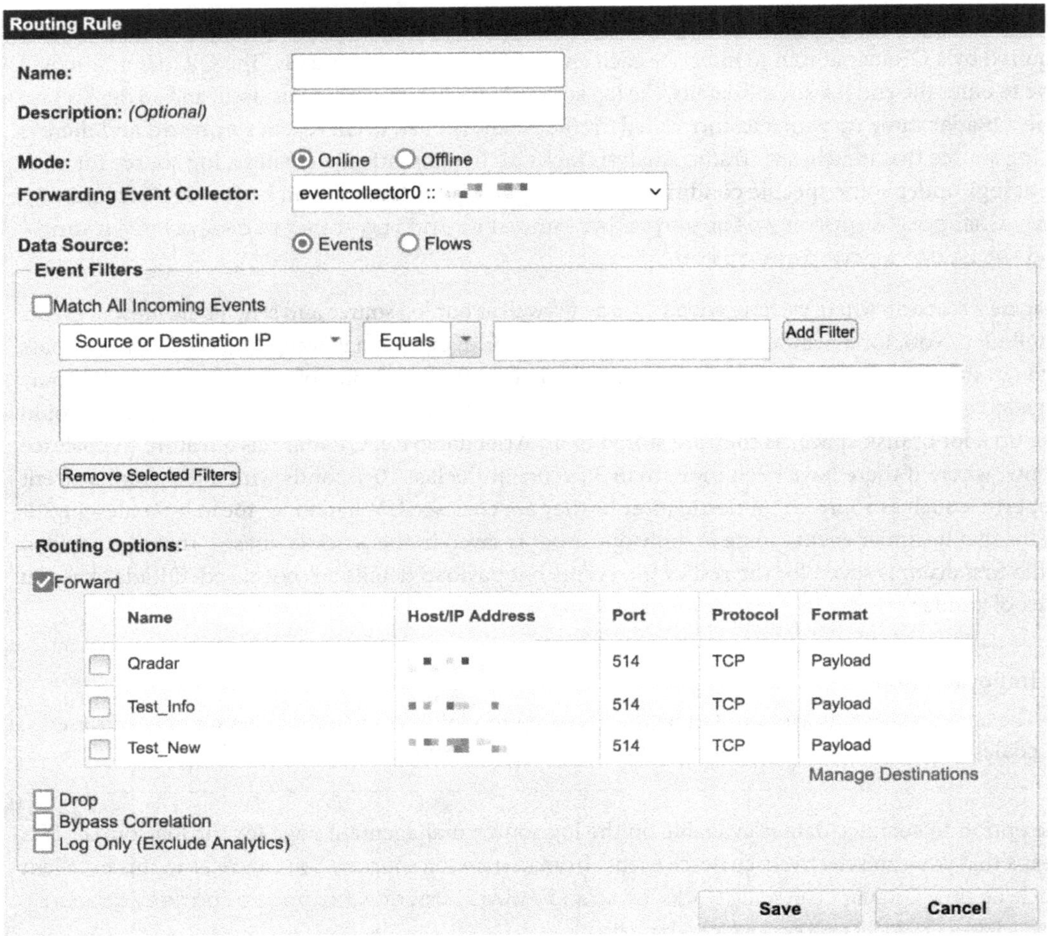

Figure 4.3 – A screenshot of the routing rules that are used for event forwarding

Note that when we set up events to be dropped using event forwarding, the number of events dropped at an instance of time is added to the license for the next instance. This means that when we drop events, these events are not counted against the license module over a period of time. This is also known as license giveback.

Another aspect of event forwarding is whether it is online forwarding or offline forwarding:

- **Online forwarding**: From the event pipeline, the event is directly sent to the destination deployment. If the destination deployment is down or there is a connection issue, the events are not forwarded to the destination.

- **Offline forwarding**: The event is first stored in the Ariel database and then the event is sent to the forwarding destination. This is used when the connection between the current deployment, from where the event is forwarded to the destination deployment, is not stable. Offline forwarding ensures that the event is forwarded to the destination. When the events are not being sent to the destination, they are buffered on the host machine.

Once the `ecs-ec` completes the processing of the event, the event is then sent to the `ecs-ep` process, where the first component is the **custom rules engine** (**CRE**). The CRE is responsible for running rules and matching each incoming event with the defined rules in the system. Along with rules, it is also responsible for matching the building blocks, which are nothing but subsets of rules. So, in a rule, multiple building blocks can be used. This provides modularity and clarity while designing and implementing rules in QRadar. Also, building blocks can be reused in different rule. These rules are responsible for generating offenses when incoming events match the rule conditions.

We will learn in detail the different types of rules when we discuss rule optimization in *Chapter 7*.

On the **Log Activity** tab in QRadar, you can see an option to live-stream logs in real time. `ecs-ep` is responsible for live-streaming events when a log activity QRadar admin requests real-time logs.

Finally, the processed events are stored in the Ariel database by the `ecs-ep` service. The Ariel database is a proprietary database to store events and flows in QRadar. It has a hierarchical structure, depending on the date and time at which the event is stored.

You may recall that the start time in event details is the time when an event hits the `ecs-ec-ingress` service, while the stored time in event details is the time when the event is stored in the Ariel database. If there is a considerable difference between the start time and store time, it means that QRadar either takes a lot of time parsing events, the EPS exceeds the license, or the rules might be too complicated.

This is where QRadar performance issues crop up. We will cover in detail how to troubleshoot performance issues in *Chapter 12*.

Understanding flows and types of flows

In the earlier chapters, we learned that flows represent the metadata for a session between two hosts. The session can span from a few seconds to minutes to hours. So, unlike an event, which is a set of complete action in itself, a flow record has data collected over a period of time. There can be a number of flows that have the same *first packet time* and *last packet time* if the flow captures the session, which lasts for more than a minute.

Flow data along with event data offers irrefutable evidence that cybercrime has taken place. QRadar is a leading security solution that has integrated flows since its inception. Over the years, new types of flow sources have been integrated into QRadar. Fundamentally, there are two different types of flow sources – internal and external. Let's look at them in detail.

Internal flow sources

Just as event collectors have the `ecs-ec-ingress` service to collect logs and send them to `ecs-ec` for parsing and processing, QRadar has the QFlow service, which can receive network traffic and convert the network data into flow data. In any organization, switches and routers are the network devices used to connect different networks or devices. Switches and routers have a port called a SPAN port that mirrors all incoming and outgoing traffic on the switch or router.

The QFlow service is responsible for collecting and analyzing flows. The first 64 bits of the payload are saved. Complete payloads are not saved, as they would easily fill up the QRadar disk space. Normalized data has different parameters, such as the following:

- The protocol
- The application
- The first packet time
- The last packet time
- A low-level category (this is determined by QFlow)
- The flow direction
- The source IP
- The destination IP
- The source port
- The destination port
- The flow source type

This flow data is then passed to the `ecs-ec` service, which is responsible for normalizing data as per configuration. There are different processes for normalizing flow data, such as deduplication, asymmetric recombination, and licensing. We will discuss these concepts one by one.

Deduplicating flows

Figure 4.4 imagines two different networks. In each network, there is a switch that is used for networking. This switch has a SPAN port, and the network data is captured by a Flow Collector. Now, both switches send network data to the Flow Collector.

Imagine that node **A** on a network communicates with node **B** on another network. In such a case, both Flow Collectors actually capture the same network data twice – once from the first SPAN port connected to **Flow Collector 1** and then from the second SPAN port connected to **Flow Collector 2**.

QRadar has a feature where the `ecs-ec` Flow Collector is capable of understanding the duplicate information and removing it. This helps simplify the data and also saves disk space.

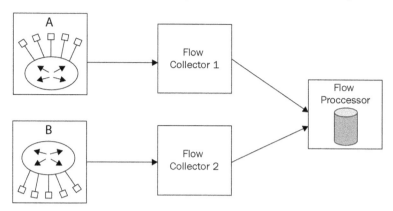

Figure 4.4 – The deduplication of flows

While configuring the Flow Collector, there are options to enable deduplication of flows. Some of the parameters related are as follows:

- Removing duplicate flows
- The external flow deduplication method
- The external flow record comparison mask

Asymmetric recombination

Refer to *Figure 4.5*, where we see a client interacting with a web server. The routing is configured in such a way that it is not necessary for the response from the web server to take the same route as the request to the web server. Because of such a peculiar scenario, the routing is called *dynamic* or *asymmetric*. The routers involved in this communication can send the flow data to QRadar. The Flow Collector will ideally receive just one part of the communication. To make sense of flow data in such scenarios, there is a concept called the asymmetric recombination of traffic.

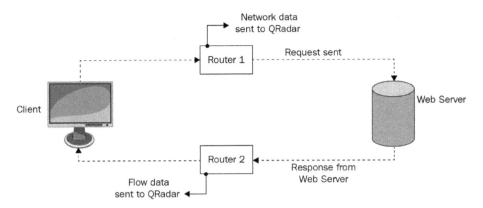

Figure 4.5 – The asymmetric recombination of flow data

This feature helps us to understand the communication so that we can design better flow rules around the flow data.

Licensing

Events are measured in EPS. Similarly, flows are measured as **Flows per Minute (FPM)**. A license will have a defined EPS and FPM for a deployment. When the license is installed, the FPM values are part of the license pool. From the license pool, FPM can be distributed to different flow processors. If the number of FPM is more than the licensed value for a considerable amount of time, then a notification is sent to the console that the flows have exceeded the license count.

A detailed description of the FPM license can be found here: https://www.ibm.com/support/pages/qradar-licenses-and-flow-data-faq.

Now that we understand what internal flow sources are, let us discuss external flow sources.

External flow sources

The way in which QFlow extracts information from the network packets, other vendors have developed different ways to fetch information from the network data. For example, CISCO has developed a proprietary way in which it collects network information from network packet data. This type of data from CISCO is called NetFlow. Similarly, other vendors have developed different technologies to monitor network data.

QRadar can ingest this proprietary network data and use it for correlation. The information received from the external flow sources can be fine-grained, helping us to create better rules around them. Another advantage to consuming data from external flow sources is that the normalization of the flow data is already done on the network device from where we collect the flows. For example, with NetFlows, the normalization is already done on the CISCO firewall, and hence, it saves a lot of CPU cycles on our flow collector or flow processor.

The following are the different types of external flow sources supported by QRadar:

- **NetFlow**: As seen in previous examples, this is one of the most popular external flow sources. This is proprietary flow information from CISCO routers and switches. There is a well-defined NetFlow template, according to this template the data is collected. If a greater number of parameters are added, fine-grained information can be collected. There are different types of NetFlows too. QRadar supports types 1, 5, 7, and 9.

- **Internet Protocol Flow Information Export (IPFIX)**: This is like NetFlow, in that IPFIX can be used to capture flow data on switches and routers. However, IPFIX can be used even by an **Intrusion Prevention System (IPS)** to send data to QRadar. The main difference between IPFIX and NetFlow is the granularity of data that is captured.

- **sFlow**: sFlow is used by multiple vendors. This is actually a sampling technique where some sample application network data is picked up, so this does not capture all the network packet information present. A sFlow source can be configured on QRadar to receive sFlow data for sFlow versions 2, 4, and 5.

- **J-Flow**: J-Flow is a proprietary flow data from Juniper Networks. It captures flow data from Juniper appliances and sends data in the **User Datagram Protocol (UDP)** format.

- **Packeteer**: This is a type of device that collects network performance data. This network performance data can then be sent to QRadar. The data sent from Packeteer to QRadar is via the UDP, which is connection-less. Therefore, there is no guarantee of data reaching QRadar.

After understanding the concept of internal and external flow sources, we must discuss the concept of superflows and their types. Superflows are a way to categorize flows that makes it easier when designing a use case or rules about flow data. Let's look at this in more detail.

Superflows and their types

With events, we have seen that if similar ones are sent to QRadar, there is a feature called coalescing that replaces them with a single event, thus saving a lot of disk space on QRadar. Similarly, with flows, when similar ones are observed by QRadar, superflows are created. A superflow is a set of similar flows. This group of flows is just counted as one flow when we count it against the FPM license.

So, fundamentally, the coalescing of events stores just one event, but superflows have all the flows in store. Superflow configuration can be done by changing parameter values, such as *create superflows*. Whenever we discuss flows, we are discussing communication between two or more hosts. So, the direction of the flow – that is, who has initiated the communication – becomes important. Based on the direction of communication and the number of hosts being interacted with, superflows are further classified into three major types:

- **Network scan**: A network scan is done to understand which ports are opened on which hosts. Usually, we use tools like nmap to perform such scans. Here, we use the ping sweep feature of nmap. This scan is like unidirectional traffic from one host, where the scan is initiated to multiple hosts in a network. This type of flow, where the source IP is the same but there are multiple destination IP addresses, is referred to as superflow type A.

- **Distributed Denial of Service (DDoS)**: In another scenario, there could be multiple hosts connecting to one host. In a typical DDoS attack, multiple hosts or, say, bots try to access a service on one host. It is typically web servers or database servers that are targeted in DDoS attacks. This is referred to as superflow type B.

- **Port scan**: The last type of superflow is where the source IP and destination IP addresses remain the same, but the source ports and destination ports may change. This is usually done to understand what services are running on the destination host. The ports that are open can suggest the type of services running on the destination host. This is referred to as superflow type C.

All these superflow types are used as conditional parameters in rules to fine-tune the QRadar rules. This also makes rules easily readable and editable. Flow data has different types of sources, such as internal and external flow sources. Under special circumstances, as discussed in this section, flows can become superflows.

In the next section, we will discuss a unique use case to ingest event data in QRadar.

Getting to know DLC

So far we have seen that to ingest data into QRadar, we need QRadar software to be installed on the VM or bare-metal servers, to have QRadar instances in the cloud, or to have QRadar appliances. But what if a customer does not want to invest in installing QRadar collector software or does not want to have an event collector? Can we still ingest events into QRadar? And if so, how?

Some customers may not have the bandwidth to procure QRadar boxes or maintain them. In this case, they can install an event collection function on a Linux machine that they already have running. This is not a dedicated event collector but a Linux box, which is installed with free software from IBM and is able to send a limited number of logs to QRadar deployment. The customer definitely needs the QRadar console in this case, but they do not require a separate event collector to collect events.

Some of the very high-security networks do not allow any inbound traffic. In such networks, DLC plays the perfect role in sending unidirectional UDP packets from the high-security networks to QRadar deployment.

Here's an example – the industrial control systems on remote islands that generate security data. Because of low bandwidths or the unavailability of the network for days, this data cannot be directly fed into QRadar. For such scenarios, DLC plays a perfect role in capturing the data and then sending over the data when the network is available. So, consider DLC as a warehouse for data till the ship (network availability) comes to pick up the cargo (data).

This DLC solution supports both UDP as well as TLS connections. At the time of writing, the DLC is configured to collect syslog data. Also, other log sources, such as Oracle, Apache, and Cisco, can send data to DLC, which, in turn, will collect data and send it over to QRadar deployment. This can be understood better with the following diagram:

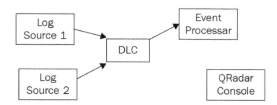

Figure 4.6 – How DLC fits into QRadar deployment

The list of supported log sources can be found here: https://www.ibm.com/docs/en/qsip/7.5?topic=collector-overview-disconnected-log.

As mentioned, DLC is free software provided by IBM and can be found on IBM Fix Central. This free software is based on Java, and hence, you will need to install the IBM SDK Java. DLC is not a managed host, and hence, the version of DLC will not match that of QRadar deployment. Also, the DLC upgrade will be separate from how we upgrade the QRadar Console and other managed hosts. You can define which logs to be collected and the maximum EPS for DLC. Invariably, when these events are sent to QRadar deployment, they will have a license throttle on the event processor or console, where the events will be counted against the license deployed on QRadar. Another point to consider is that traffic can be encapsulated and encrypted when it is sent from DLC to the processor.

In short, DLC is a good solution for customers who do not want to invest heavily in buying QRadar appliances or maintaining them. Conversely, it is not a managed host.

Summary

In this chapter, we have looked at all the different ways data is ingested in QRadar. We also discussed different use cases and how QRadar provides solutions such as DLC, superflows, and event forwarding. As compliance and requirements change over time, QRadar comes up with unique use cases to comply with and fulfill requirements.

In the next chapter, we will look at how QRadar ensures that no data is lost while collecting and processing. We will also learn how to integrate data that is not supported out of the box.

5
Leaving No Data Behind

In the previous chapter, we learned how event data is collected and consumed by QRadar. We learned that protocols are needed to collect data while **Device Support Modules** (**DSMs**) are required to parse data. Consider a scenario where we want to ingest event data into QRadar but there is no supported DSM. The first thing is to know what the supported DSMs are.

Every month, IBM releases a new DSM guide, a document on how to integrate log sources with QRadar. If your log source is not a part of this DSM guide, then the event data ingested is either categorized as *Stored* or *Unknown*. The event data is not parsed. That does not help us with correlation when it comes to matching events with rules. So, whatever event data we are ingesting in QRadar should be parsed properly.

Using a tool called **DSM Editor**, we can create custom parsers for any type of data that is ingested in QRadar. In this chapter, we will discuss the steps to create custom parsers in QRadar in detail. Along with DSM Editor, there are a few other concepts that we will discuss that will help us understand how QRadar ensures that no data is left behind.

In a nutshell, we will cover the following topics in this chapter:

- Queues and buffers
- DSM Editor

Understanding queues and buffers

We discussed in the previous chapter how if the number of events being ingested in QRadar is more than the license threshold, a system notification is sent by the console on the UI. Let us dig deeper to discover how events are managed in different scenarios.

Persistent queues

QRadar changed its design concept to introduce persistent queues. This was primarily done to avoid event loss. We understand that there are three basic services in the event pipeline:

- `ecs-ec-ingress`
- `ecs-ec`
- `ecs-ep`

Once the events hit the event pipeline, QRadar ensures that they are ingested successfully.

Imagine that the `ecs-ec` service has crashed. What will happen to the incoming events? `ecs-ec-ingress` will still be collecting events and trying to send them to `ecs-ec` for parsing. If the `ecs-ec` service is down, then the events coming in are stored temporarily in the ingress persistent queue. Once the `ecs-ec` service is up and running again, the events can be pulled from the ingress persistent queue. This happens on a first-in-first-out basis.

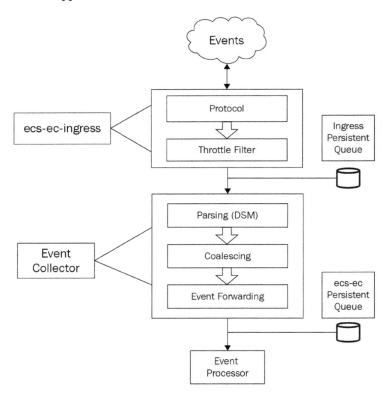

Figure 5.1 – Persistent queues

Along the same lines, if the event processor service is down, that is, if `ecs-ep` is down, then `ecs-ec` will send the events to the `ecs-ec` persistent queue. Once the `ecs-ep` service is up, the events will be picked up from the `ecs-ec` persistent queue.

Both queues can grow as large as the `/store` partition on which they are configured. But ideally, the size of the queue should be close to zero.

The mechanism of how any persistent queue works is configured by default and no human intervention is required.

The locations of the queues are as follows:

- Ingress persistent queue: `/store/persistent_queue/ecs-ec-ingress.ecs-ec-ingress`

- The `ecs-ec` persistent queue: `/store/persistent_queue/ecs-ec.ecs-ec`

> **Important note**
> If there are disk-full issues or other issues because of persistent queues, contact IBM Support. Do not edit or copy and paste the files from persistent queues to any other location.

We have discussed persistent queues. In the next section, we will cover in-memory queues and disk buffers.

In-memory queues and disk buffers

The other design feature in QRadar is the in-memory queue. From earlier chapters, we have understood that `ecs-ec-ingress` uses different types of protocols to pull or accept different kinds of event logs. For each of the protocols supported by QRadar, there are in-memory queues allocated. For example, say there is enough memory to handle 10,000 syslog events; there will also be enough memory for 10,000 JDBC events, and so on. This is configured by default.

In-memory queues help to store events temporarily before they are pushed to the licensing module, where the events are counted against the EPS license.

If the number of events is more than the license, then the events are pushed to the disk buffer. This disk buffer acts as temporary storage until the incoming EPS falls back. Please note that this mechanism of disk buffer is useful for small bursts of events. If the incoming EPS remains consistently high, a system notification is sent to the console and the events are then dropped.

In *Figure 5.2*, you can see the in-memory protocol queues and the disk buffer, which is in between the protocol module and the licensing module in the `ecs-ec-ingress` queue.

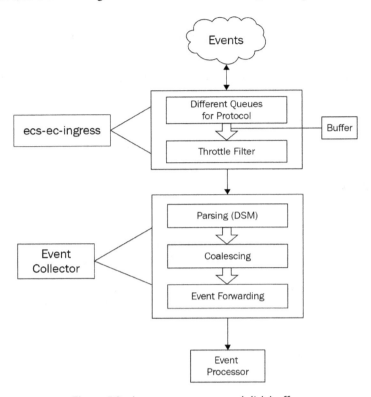

Figure 5.2 – In-memory queue and disk buffer

A very well-explained note on how burst handling is done in QRadar can be found here: `https://www.ibm.com/support/pages/qradar-event-and-flow-burst-handling-buffer`.

> **Important note**
>
> To better understand all the concepts described in this chapter and the ones before, do refer to the figures provided in this book.

The fundamentals remain the same across all versions of QRadar. These design changes of introducing persistent queues, in-memory queues, and disk buffers have been game changers when it comes to addressing the challenge of events being dropped. In the next section, we will discuss how to parse unsupported event data.

Getting to know DSM Editor

We have discussed in detail event data and the ingestion of event data in QRadar and now understand that IBM provides DSMs out of the box for QRadar to parse incoming event data. What happens when IBM does not have a DSM for a data source that you want to ingest? What would be the state of the ingested event data? Will it be partially parsed? To answer these queries, IBM has introduced a tool called DSM Editor. DSM Editor is built into Qradar; no special package is required for its installation.

In earlier versions of Qradar, there was a feature called Universal Log Source, where we had to define the parsing logic for incoming data. Parsers had to be written. As it was a manual process, it was not very efficient. But now with the DSM Editor, most of the processing is automatic.

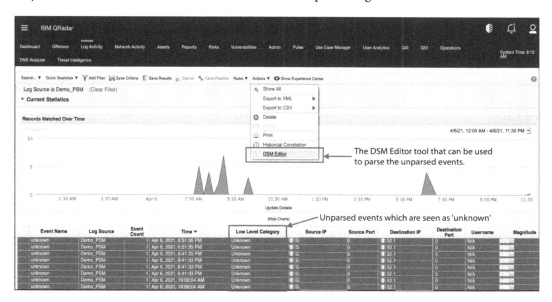

Figure 5.3 – DSM Editor – unparsed events

In the preceding figure, we can see a few events that have a **Low Level Category** value of **Unknown** from a demo log source. This demo log source does not have a known DSM installed in QRadar.

> **Important note**
>
> You may see an event's **Low Level Category** value appear as **Stored** or **Unknown**. Let us learn more about that:
>
> **Stored**: When QRadar is able to associate an event to a log source but is not able to parse the event, then that event is labeled **Stored**. For example, Blue Coat Web Security is a supported end device, and hence QRadar has a DSM available for this. But if the Blue Coat Web Security appliance is upgraded, there is a possibility that a new type of event is also included in the upgrade. Now QRadar will be able to associate the new event coming from the upgraded appliance but may not be able to parse it because of the different regex. In such scenarios, the new event type will be labeled **Stored**.
>
> **Unknown**: We know that when an event is parsed, it is also mapped to the known QID. If the mapping of a particular type of event is not present, then such an event will be labelled **Unknown** for **Low Level Category**. Taking the last example, if we make changes to the DSM logic to add a new type of event but do not map the new type of event, then it will be an **Unknown** event.
>
> When DSMs are updated, QID mapping is also updated to accommodate the changes.

DSM Editor picks up the complete event payload and tries to parse it as per the configuration. Event data is always bound by a certain event format. It depends on the end application that generates the event data. QRadar has the feature to autodetect all the properties of the event data if the event data is in any of the following formats:

- JSON
- XML
- CEF
- LEEF
- Name-value pair

For example, a new application for intrusion prevention generates logs in JSON format. But QRadar does not have a DSM defined for this intrusion prevention application. As the logs from this new application are in JSON format, QRadar can autodetect the events and parse the events accordingly. No manual intervention is required. For this feature, you need to enable it in the **Configuration** tab of DSM Editor.

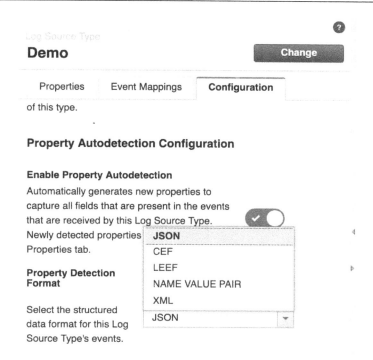

Figure 5.4 – DSM Editor – autodetection option

Now let us imagine that the event formats that are not autodetected and do not have DSMs. For these events, each event type can be broken down into multiple properties. These event properties can be defined using regexes.

> **What is a regex?**
>
> Regexes are regular expressions that are used programmatically to represent a number of different permutations and combinations of alpha-numeric characters.

Once the properties are defined, it is time to map the events. Events can be mapped using the event ID and event category. The event ID is the primary source used to map an unknown event to a known event.

For example, our intrusion prevention application created an event for denying access to a particular URL that was being accessed. From the event data, we understand that this is like an *Access Denied* event, which is a common event in, say, a CISCO firewall. As the end goal of the event is the same, we can map the new intrusion prevention event to an event called *Access Denied*. Now, the *Access Denied* event for our new intrusion prevention application is separate from the one defined for the CISCO firewall. They will have different QIDs, that is, event IDs.

After the event is mapped once, all upcoming events with similar payloads will be parsed as *Access Denied* events. This helps to automate event parsing.

A detailed discussion of how event property configuration can be done using DSM Editor can be found here:

```
https://www.ibm.com/docs/en/qsip/7.5?topic=qradar-property-
configuration-in-dsm-editor
```

Using the same example, there is another scenario that we need to discuss. What if the intrusion prevention application was installed on 25 different machines? In this case, we would have to create 25 different log sources and then parse each one using DSM Editor. This would be very inefficient and time consuming. QRadar has a feature where you can define the log sources that are yet to send logs and also make sure that the logs are parsed correctly.

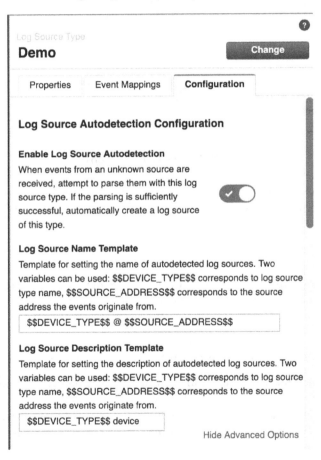

Figure 5.5 – Log source detection using DSM Editor

In *Figure 5.6*, we see that we have defined the log source using DEVICE_TYPE and SOURCE_ADDRESS. So, any intrusion prevention logs coming from the 24 other machines will create a separate log source automatically. The log source name as defined here would be cali_ips@9.9.9.9, where cali_ips could be the device type and the source address could be the source IP picked up from the event payload.

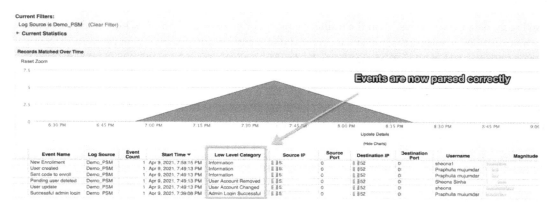

Figure 5.6 – Parsed logs – DSM Editor

Finally, after configuring DSM Editor, the end result would be well-parsed logs. If *Figure 5.3* depicts the state of the logs before using DSM Editor, *Figure 5.7* shows how the parsed logs would look afterward. In this way, DSM Editor is an automated tool that can be used to parse any type of event data.

Let us now look at a very practical example of how DSM Editor is used. Till now, we have discussed collecting events and flows from systems that are primarily software-based and dealing with security in one way or another. For example, we have seen how to collect logs from firewalls, flows from switches, and log data from endpoint agents such as Windows.

But there are other programmable systems and devices that interact with the physical environment. For example, in healthcare, there are so many devices that have internet connectivity. These devices may collect health-based information from patients, process the data, and save it somewhere on their servers. Just like security devices, such systems should be able to use the capabilities of our QRadar SIEM. This doesn't just go for healthcare but also industrial control systems, transportation systems, building automation systems, and more. This category of programmable systems and devices is known as **Operational Technology (OT)**, and the security designed around it is called OT security.

The **National Institute of Standards and Technology (NIST)** is a US-based institute responsible for developing standards in the field of cybersecurity. NIST has developed a framework that recommends that OT systems send data to SIEM systems. This is documented in *Guide to Operational Technology (OT) Security*, which you can refer to here: https://csrc.nist.gov/News/2022/guide-to-operational-technology-ot-security.

In such scenarios, the security data generated by healthcare devices or industrial automation devices is then sent to the SIEM system, which in our case is QRadar. DSM Editor plays a pivotal role in making sense of this data by parsing the logs. Further rules can be written for the log sources of OT devices.

Summary

Leaving no data behind is based on the military concept of *no man left behind*. With QRadar, we have taken utmost care to ensure that no data is dropped under any circumstances (except more events being sent to QRadar than the licensed amount). To do this, we have introduced the concepts of queues, buffers, and so on. We have also created a tool called DSM Editor to parse unsupported log sources. This chapter will help you ingest any types of logs that are generated in your environment. Whether these logs are based on applications, custom operating systems, or **Internet of Things (IoT)** devices, you will be well aware of how to use log data for ingestion.

In the next chapter, we will talk more about data and how it is to be searched. We will cover the fundamentals of searches and how to optimize searches.

6

QRadar Searches

If you ever take a philosophical view on the technical aspect of **Security Information and Event Management (SIEM)**, a fleeting thought is that SIEM is nothing but a very advanced search. Everything else is built to collect data, run correlation, run analytics, and then display the search in a better way. That is kind of true.

What we have learned and discussed so far is the way QRadar is deployed, its different components, how data is ingested, and so on. Now, we will start with using the ingested data to make sense of it. QRadar search is the most fundamental feature of QRadar. In this chapter, we will discuss QRadar searches and how we can optimize them and use them to the best that they can offer.

The following topics will be covered in this chapter:

- How do searches work?
- QRadar services involved in searches
- Different types of QRadar searches
- Concept of data accumulation
- Different ways to tune QRadar searches

How do searches work?

Though we have briefly discussed the idea of searching, let us dig deep into it and understand the mechanism of QRadar search.

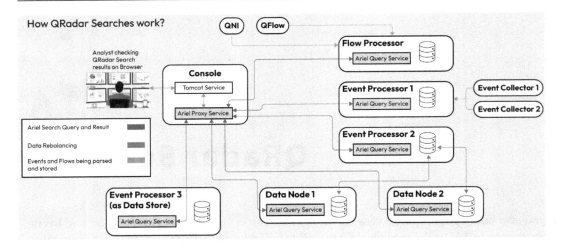

Figure 6.1 – Components involved in a QRadar search

In the preceding figure, we have tried to cover all the QRadar components that are involved in a QRadar search. We can see a security analyst on the left-hand side trying to run a search on the QRadar Console **Graphical User Interface (GUI)**. We can also see three Event Processors and one Flow Processor where data is stored. Then there are Data Node 1 and Data Node 2, which are attached to Event Processor 2. There are two Event Collectors, which are collecting logs, and those logs are stored on Event Processor 1. Similarly, flows are collected by **QRadar Network Insights (QNI)** and another Flow Collector (QFlow service) and are sent to the Flow Processor.

In the figure, we can see legends that are defined. The blue colored line depicts the search query and search result, and the orange line depicts data rebalancing among Event Processor 2, Data Node 1, and Data Node 2. The green line depicts the event and flow collection and processing.

To simplify that further, consider a scenario where an analyst wants to search for events from logs from the last 5 minutes. The analyst logs on to the QRadar Console and navigates to the **Log Activity** tab. Then, they execute the search using the filters available.

Now, with a basic understanding of how searches work, let us discuss the different aspects of a QRadar search.

Services involved in a QRadar search

QRadar searches, if not designed well, can lead to performance issues on QRadar. The various symptoms that we can see (not necessarily only because of bad search) are as follows:

- Searches become slow

- Reports take a lot of time to complete (especially reports that are run manually on raw data)

- The QRadar UI becomes slow

- QRadar partition for /transient may become full

Apart from these, however, there could be many more issues that could arise. To resolve such issues, you should know which services need to be restarted (if ever), and this section is intended to make you aware of these very services. Let's get started!

In *Chapter 1*, we discussed all the major services in QRadar. Out of those, the following are the services that are involved when running searches:

- **Tomcat service**: Whenever you want to create or run a search, open the QRadar Console and go to **Log Activity**. The QRadar UI is controlled by the Tomcat service, and so is the **Log Activity** tab. So, when an analyst tries to create a search, the Tomcat service accepts the request and sends the data to the Ariel proxy server service. The other function of the Tomcat service is to display the queried data to the analyst. The graphs and pie charts are all controlled by the Tomcat service.

- **Ariel proxy server service**: This search is a query that is created by the user, and the Ariel proxy server sends this query to all the components that run the Ariel query service. It is also responsible for aggregating the search results from all the QRadar components that have the Ariel query server service.

- **Ariel query server service**: The request from the Ariel proxy server service is received by the Ariel query service, which, in turn, queries the Ariel database that resides locally on Processor and Data Nodes (if any) box. The result of the query is then sent by the Ariel query server service to the Ariel proxy server.

From *Chapter 1*, we understand the function of the Data Node. Let's discuss the chronology of search when a Data Node is also installed. Let's say we need data from all the F5 firewall devices deployed for 15 minutes when access was denied to certain URLs.

So, in the search, we add filters and create the search query. Tomcat sends this customized query to the Ariel proxy server. The Ariel proxy server knows all the managed hosts where the Ariel data is stored. In this case, Ariel data is also present on Data Nodes. In *Chapter 1*, we learned how data is balanced between the Data Nodes and processors. So, the search request reaches all the processors and Data Nodes. Processors and Data Nodes run the Ariel query service, which queries individual Ariel databases and sends the data back to the Ariel proxy server service. The thing to note here is that there are parallel connections set from the Console to all the managed hosts hosting the Ariel database. This improves the search performance, as the same data is fetched by a greater number of threads on different managed hosts.

> **Important note**
> In huge environments where multiple searches are running in parallel, it is advisable to use Data Nodes.

Now we understand the chronology of the QRadar search, let us move on and discuss the different types of searches with examples.

Different types of QRadar searches

QRadar admins and analysts all use searches in different ways. As such, there are many ways in which searches can be customized and reused. Searches can even be assigned to different users, set as default, or included in a dashboard. In the following subsections, we look at the three types of searches available.

Default searches

QRadar provides some out-of-the-box searches, which can be accessed by clicking on the **Quick Searches** drop-down arrow shown in the following screenshot. It is highly recommended to use these default searches, as the performance of these searches is much more optimized.

Figure 6.2 – Quick searches option

They also cover a large spectrum of different searches, such as top 10 events, **events per second** (**EPS**), **flows per minute** (**FPM**), offense searches, and so on. When you click on the **Quick Searches** button, you should be able to see the available searches, as shown in the following screenshot:

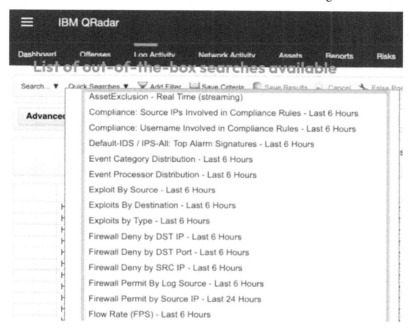

Figure 6.3 – Out-of-the-box searches under Quick Searches

Apart from default searches, you also have the option to customize a search as per your business's requirements.

Customized searches

To create a search as per your requirements, click on the **Search** drop-down box and click on **New Search**. You will be presented with the following three options:

- **Saved Searches**: These searches include the out-of-the-box searches as well as searches created and saved by any other QRadar system user. You can directly select any of the saved searches if they match your requirements or are close to your requirements. Once selected, you will need to load the search. Once loaded, you can see the details of the filters as well as how the search is grouped. In this instance, if the requirements are different, you can make changes to **Search Parameters**, **Column Definition**, or **Time Range**. Once the changes are done, you can hit the **Search** button. The actual search is loaded. If you are not using **Saved Searches**, you can directly use **Search Parameters** as filters for the search, **Column Definition** as the content of the search, and **Time Range** as the period of the search.

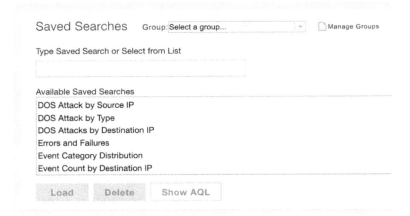

Figure 6.4 – Available Saved Searches

If QRadar users create more searches and save them, you will see those searches in the **Saved Searches** section.

- **Basic Search**: Other than saved searches, if you need to create a brand-new search based on the different filters available, you should choose the **Basic Search** option. You can determine **Time Range** using the **Real Time (streaming)**, **Last Interval (auto fresh)**, **Recent**, and **Specific Interval** options. Then, you have the option to select the columns to be displayed for the search. Finally, at the bottom, we have the option to select **Search Parameters** on which we will set the filters. You can view these components in the following screenshot:

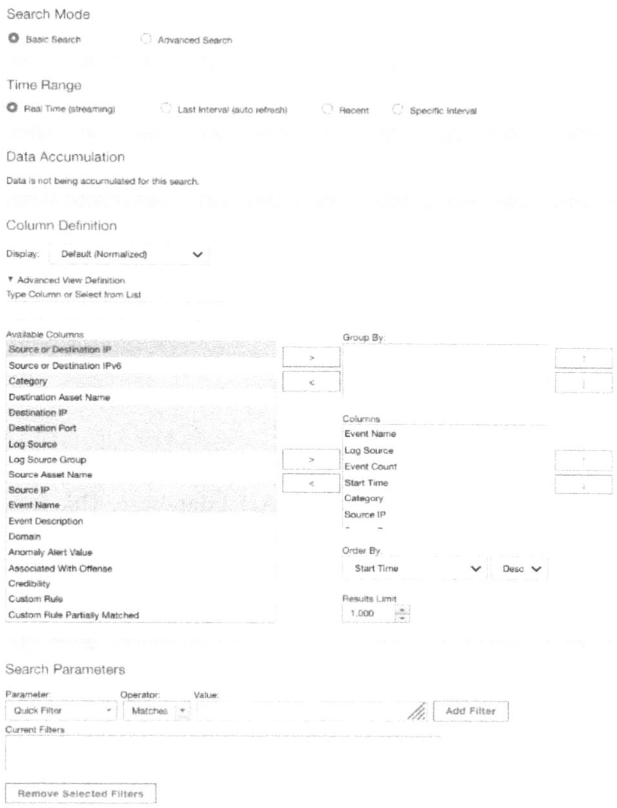

Search Mode

○ Basic Search   ○ Advanced Search

Time Range

○ Real Time (streaming)   ○ Last Interval (auto refresh)   ○ Recent   ○ Specific Interval

Data Accumulation

Data is not being accumulated for this search.

Column Definition

Display: Default (Normalized)   ∨

▼ Advanced View Definition
Type Column or Select from List

Available Columns   Group By:
Source or Destination IP
Source or Destination IPv6   >
Category   <
Destination Asset Name
Destination IP   Columns
Destination Port   Event Name
Log Source   Log Source
Log Source Group   >   Event Count
Source Asset Name   Start Time
Source IP   <   Category
Event Name   Source IP
Event Description   -   -
Domain
Anomaly Alert Value   Order By:
Associated With Offense   Start Time   ∨   Desc ∨
Credibility
Custom Rule   Results Limit
Custom Rule Partially Matched   1,000   ⇅

Search Parameters

Parameter: Operator: Value:
Quick Filter   ▾   Matches   ▾   ⁄⁄   Add Filter
Current Filters

Remove Selected Filters

Figure 6.5 – Basic Search options

In the *QRadar search tuning* section, we have a subsection for how to choose the filters for a search. When creating a basic search, do go through the best practices mentioned there.

- **Advanced Search**: When you select the option of **Advanced Search**, you will notice that you no longer see **Search Parameters**, **Column Definition**, or **Time Range**. All you see is a blank box. So, you might wonder how to use **Advanced Search**.

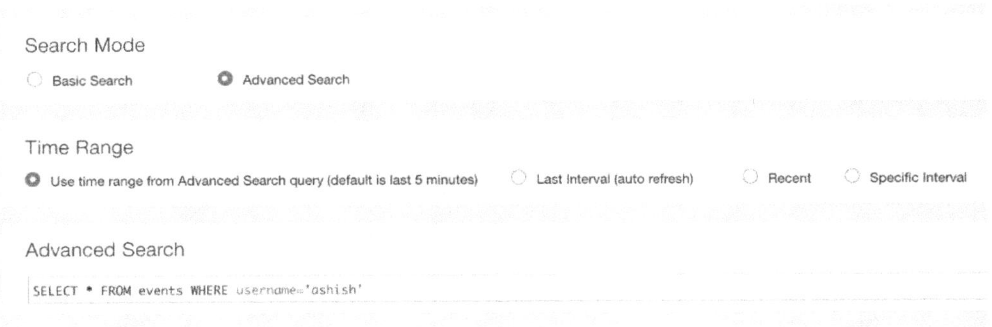

Search Mode

○ Basic Search   ◉ Advanced Search

Time Range

◉ Use time range from Advanced Search query (default is last 5 minutes)   ○ Last Interval (auto refresh)   ○ Recent   ○ Specific Interval

Advanced Search

```
SELECT * FROM events WHERE username='ashish'
```

Figure 6.6 – The Advanced Search option in QRadar

More about Advanced Search

The database that stores Events and Flows in QRadar is known as the Ariel database. This is a proprietary database created by the IBM QRadar team. So, to query this proprietary database, IBM has also created a query language, which is known as **Ariel Query Language** (**AQL**). For the most part, it works very much like **Structured Query Language** (**SQL**), with keywords such as SELECT, FROM, WHERE, and so on. But AQL also has a few more keywords. There is an excellent description of the keywords and structure, and actual sample examples here: https://www.ibm.com/docs/en/qsip/7.5?topic=aql-ariel-query-language-in-qradar-user-interface.

AQL queries may get complex when a large number of conditions are used. These queries can be used to get pinpoint datasets for reports and dashboards. AQL queries can also include reference data queries. Here is an example of an AQL query:

```
SELECT sourceip, COUNT(*) as events_count
FROM events
WHERE severity >= 7 AND destinationip = '192.168.1.10'
GROUP BY sourceip
HAVING events_count > 100
ORDER BY events_count DESC
LAST 15 minutes
```

The query displays the source IP address and the number of events, which is depicted by events_count. This data is being queried from the events database (there is also a flows database available, as AQL can query assets and vulnerabilities, too).

Then we have conditions for the severity of the event and destination IP address. The GROUP BY clause helps to collate events based on the source IP address, as it would give us the right count. Another condition we have for the display of the event is that the source IP should have more than 100 events (as the event count). For those source IPs whose event count is less than 100, the source IP address will not be displayed.

The ORDER BY clause helps to sort the events based on the parameter mentioned. In this example, we have event count as the parameter and so the source IP with the maximum event count will be displayed first. To end this query, we do not want to query the complete events database but we would like to look at events from the last 15 minutes only.

Now, let's move on to discuss quick filter.

Searches using quick filter

In short, a quick filter search is like a Google-type search in QRadar. You can directly mention the parameter that you are searching for. For example, you can directly enter an IP address, username, MD5 hash, and so on. If you input an IP address, the search result may have the IP address but it could either be a source IP address or a destination IP address. It could also be a device IP address if that is part of the event payload.

It is one of the fastest ways to search if you know the exact terms to use for the search. It can be accessed from **Log Activity** or the **Network Activity** tab.

There is one caveat while using the quick filter search. As mentioned, the search is based on payload. When the search is run, the payload information for the time span mentioned in the search is indexed. The process of indexing may take considerable time depending on the amount of data to be indexed. So, if the payload data is already indexed, then the quick filter search becomes faster.

Now, the obvious question is, what is indexing? We will cover indexing in the *QRadar search tuning* section. But before jumping on to search tuning, it is important to first understand the concept and explanation of data accumulation. In the *QRadar search tuning* section, we will also delve into the concept of data accumulation.

Data accumulation

Data accumulation is one of the important concepts to understand. Many of the applications, such as dashboards and reports, work on data accumulation. So, what is data accumulation in QRadar?

In *Chapter 1*, we spoke about the hostcontext service and we saw that there are multiple subservices of hostcontext. One of them is the accumulator service. If configured, the accumulator service is responsible for keeping a separate copy of the data. The data from the Ariel database is not removed or edited in any way. It is just that a separate copy of the small set of data from the Ariel database is maintained. This separate copy of data is called **Global View** (**GV**) in QRadar.

For example, we have configured to accumulate Cisco firewall logs where the event name is Connection Denied. The accumulator service will pick up all the events named Connection Denied and store them in a GV. GVs are storage compartments where specific events can be saved or duplicated for easier access. The events will be picked up from all the processors/Console, wherever we have configured to process Cisco firewall logs. So, the accumulator service runs on processors/the Console and does not run on collectors or other components.

As mentioned, we are collecting the Cisco firewall logs for a particular event. If we need to create a scheduled daily report for these events, when the report has to be executed, the search will have to be run. Now, depending on the number of processors and amount of data to be queried, the search may take a long time and thus the report will take a long time. To avoid this, when the report is scheduled, a GV is created. This GV will duplicate the events that are processed for the Cisco firewall. Thus, a duplicate set of required events is already present when the report must be run. This helps in keeping the system efficient and free from unnecessary search load at the time of report generation.

Taking the example forward, if we plan to generate a daily scheduled report, a GV is created. This GV will start collecting all the processed events with the event name `Connection Denied` and store them in the GV. When the scheduled report has to run, as the GV exists, the report will collect data directly from the GV rather than searching for it in the Ariel database.

Obviously, searching for data from, say, a 2-TB /store/ariel partition (the complete ariel database) would be a lot more computationally expensive than running a search on 200 MB data from GV (note that the figures used here are symbolic and may differ in your environment). The point that we should understand here is that the searches are efficient when the right kind of data is already accumulated and segregated based on GVs.

Like scheduled reports, certain dashboards are like time-series graphs. If time-series graphs are configured, again, a GV is created for it. So, when the dashboard is accessed, it becomes more efficient to pull the requested data from the GV than pulling the same data from the Ariel database.

Another question that you might have is *Where are these GVs stored*? These GVs are stored in the /store partition. Once you schedule a report or create a time-series graph for a dashboard, you can view the GV data on the QRadar UI. You can navigate to **Admin | Aggregated Data Management**. You will see that there is a drop-down box with options for **Reports**, **Time Series**, **ADE Rules**, and **Aggregated Data View**. We will discuss **ADE Rules** in detail in *Chapter 7*.

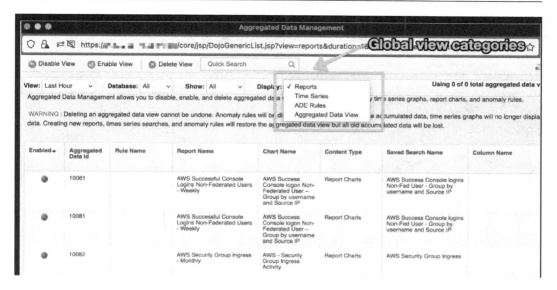

Figure 6.7 – Aggregated Data Management view

In the preceding figure, we can see the aggregated data collected over the last hour for the **Reports** section. These reports will be scheduled reports. This view is usually used to check or delete the views in case the data in the view is corrupted.

After understanding the concept of accumulated data, it is very important to note that there are limitations to the number of GVs that can be created. GVs are nothing but copies of subsets of data. So, we cannot indiscriminately create GVs. The following are the different aspects that should be considered when configuring data for accumulation:

- Whenever you create scheduled reports or time-series graphs, you should regularly check the **Aggregated Data Management** view to understand the amount of data that is collected in the GVs. If the amount of data is drastically increasing over time, clearly, the GV is probably not designed correctly. The following screenshot shows where you can check the amount of data written by a GV:

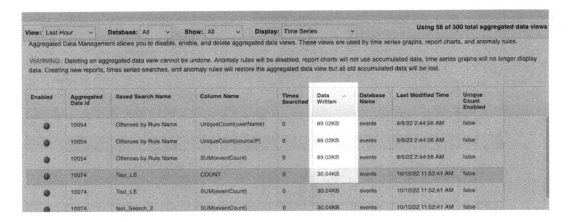

Figure 6.8 – Data Written by a GV

The highlighted part of the figure shows the data aggregation view of the GVs created because of the time-series graphs. We can see that the maximum data written is in KB, which is acceptable. Had the **Data Written** column shown data in GB, we would need to be very careful about how many such GVs were created.

- The second point is the number of GVs that are generated. You may encounter an error if too many GVs are generated. You can determine the number of GVs by using the **Aggregated Data Management** view and looking at the number of GVs generated. You can disable them or delete them, as explained in the following technote. Check this technote for a detailed explanation of the issue: https://www.ibm.com/support/pages/qradar-aggregated-data-limit-has-been-reached.

- Always, think before you act. There is a concept called **Master Aggregated Data View (MADV)**, which we will explain in detail in the last section of this chapter. You can create MADVs so that your searches can piggyback on a master GV and save a lot of space and computation for QRadar.

In this section, we understood how data accumulation works and how to use it for efficient searches. In the next section, we will explore more options to fine-tune QRadar searches.

QRadar search tuning

Earlier, we looked at the different types of searches. QRadar searches are one of the most computationally expensive functions. If done right by following a few rules, the searches will work smoothly and efficiently. Otherwise, you will end up with performance issues on various fronts. All the searches are initiated from the QRadar Console and hence, if searches are done wrong, it might affect the other functionalities or services on the Console.

Here are a few rules when it comes to QRadar searches. Let's dive in!

Indexing and index management

An index is metadata that is generated for the data in the Ariel database. This index data can be generated as soon as the events or flows are ingested in QRadar, or the index can also be generated before running searches (i.e., post-data ingestion). Indexing is used to make the QRadar searches fast and efficient.

Indexing is enabled on the property of the events/flows. For example, *source IP* would be an excellent event property to be indexed because many of our searches would include *source IP* as a filter. In the QRadar UI, we have an **Index Management** page where we can see which event properties are indexed.

On the QRadar UI, go to **Admin | Index Management**.

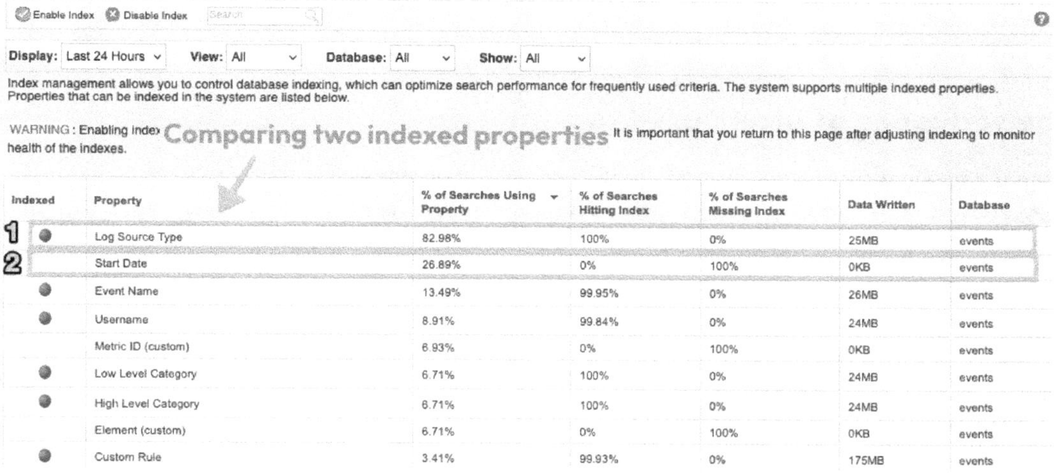

Indexed	Property	% of Searches Using Property	% of Searches Hitting Index	% of Searches Missing Index	Data Written	Database
●	Log Source Type	82.98%	100%	0%	25MB	events
	Start Date	26.89%	0%	100%	0KB	events
●	Event Name	13.49%	99.95%	0%	26MB	events
●	Username	8.91%	99.84%	0%	24MB	events
	Metric ID (custom)	8.93%	0%	100%	0KB	events
●	Low Level Category	6.71%	100%	0%	24MB	events
●	High Level Category	6.71%	100%	0%	24MB	events
	Element (custom)	6.71%	0%	100%	0KB	events
●	Custom Rule	3.41%	99.93%	0%	175MB	events

Figure 6.9 – Index Management page

Let us understand indexing using the preceding figure. The figure shows the event and flow properties used in searches for the last 24 hours. Similarly, you can have the statistics for the last hour to the last 30 days. This view gives you an idea of whether the indexing is optimized or not.

We know that an index is a kind of metadata, which means that when we create an index, some space is occupied by that index. So, you just cannot create an index for each and every event/flow property. If done, the index itself will be so large that a lot of disk space will be utilized and it would be counterproductive. So, a balance should be maintained on the number of properties that are indexed.

In *Figure 6.9*, we have highlighted two properties. For the **Log Source Type** event property, indexing is enabled. Of all the searches made in the last 24 hours, approximately 83% of searches were on the **Log Source Type** property. As indexing was enabled for this property, 100% of the searches involving **Log Source Type** were using the indexing enabled for the property.

For the **Start Date** event property, we see that of the total searches made in the last 24 hours, approximately 26.9% of the searches had the **Start Date** property. Also, this property is not enabled. If we enable indexing on **Start Date**, statistically speaking, 26.9% searches would be more efficient and faster.

So, what can we deduce from the previous example of indexing different event properties?

- You may broadly say that if the property is used in more than 30% of the searches in the recent past, then such a property should be indexed.

- If a property is not used often, say less than 30% of the total searches, then an enabled indexed property is not being used optimally. Either use the indexed property in the search or else disable the enabled indexed property.

These recommendations will help you tune your searches and will save a lot of time and computing resources while running your searches.

The sequence of filters used in a query

From the previous section, we understand the importance of indexing and how to manage indexed properties. Now, the second factor is how to design a query. There are a few rules, which we will discuss next.

Start small

Narrow the scope of your search. Time is the most important factor while searching. If you know the approximate time for which you want to search data, start your search with the lowest time frame. Let us discuss an example of how to start small.

Imagine that there is a new malware attack discovered in the wild. Currently, there are no **Indicators of Compromise (IOCs)** that match the discovered malware in QRadar deployment. There are new rules added to detect this malware and none of the rules has triggered any offense. Looking at the malware details, we recollect looking at a similar pattern of events, but those events were seen last December. In such a scenario, we have the IOCs, which would be MD5 hashes or IP addresses. As per the strategy, we should start small. So, we would take mid-December as a starting point. Initially, we will start with a 1-day search for, say, the 15th of December. If we do not find the required IOCs, we can increase the scope of the search for a week where we can start from the 12th of December to, say, the 19th of December. If we do not find any IOCs, then we increase to 2 weeks, and so on.

Now, the question is why don't we immediately search events for a complete month? The reason is that if you want to find a needle in a haystack, you should divide the haystack into smaller units and then search. This strategy helps QRadar to optimize its resources when searching for a larger timeslot when a part of that timeslot has already been completed. This also avoids unnecessary long searches, which would affect overall QRadar performance.

Use indexed properties

Previously, we discussed indexing. As per this strategy, if the search has indexed properties, use them as the next search filter. If the search does not have an indexed property, go back to the **Index Management** page and analyze the properties that we are using in the search. Check whether any of the properties can be indexed as per our recommendation in the previous section, which is the 30% rule. If yes, go ahead and index the property. Indexing will take some time depending on the amount of data to be indexed.

Narrow down the search

The filters should be used in such a way that the first filter should narrow down the search to the least number of events to be searched.

Consider an example in which we want to search for events with filters A, B, and C, and there are about 1 million events in the database. Out of these three filters, you should know which filter will narrow down your search the most. Say A was *Event Low-level Category*, B was the username, and C was the event name; we would ideally use filter C, which is for the event name. Event names are usually unique and are a more specific filter than *Event Low-level Category* or username. This will reduce the number of events to be searched by filters B and A. After the event name, the sequence should have the username as the second filter and *Event Low-Level Category* as the third filter.

The idea here is to narrow down the scope of the search using filters.

Add Data Nodes

Data Nodes are QRadar appliances that are designed to store data. Data Nodes may be attached to an Event Processor, Flow Processor, or the Console. When the events are processed by a processor or the Console, the data can be stored either on the processor/Console or a Data Node. When Data Nodes are attached, data is rebalanced between the appliances. The advantages of adding Data Nodes are as follows:

- We have more storage space for Ariel data
- Searches are efficient as the number of connections querying the same data increases

> **Important note**
> We discussed, in detail, how Data Nodes work in *Chapter 1*. You can go back and give it a re-read as a refresher if needed.

Creating a MADV

We have discussed data accumulation and we know that there is a limitation on the number of GVs that can be created. The way to overcome this limitation is the creation of a MADV. If you perform statistical analysis on the filters and the data searched, it is evident that a subset of data is searched more than the rest. For example, for firewall logs, a SOC analyst may search `Firewall Deny` events more than `Firewall Accept` events. This is exactly what we also see when we open the **Index Management** page. The design logic behind MADV is to create a view or snapshot of a limited amount of data that will be searched multiple times. When we run a normal search, the search is run on the complete Ariel database. But when we use MADV, the same search is run on a limited amount of data. This is also known as piggybacking a search.

There are a few considerations while creating a MADV:

- We first need to understand for which searches this MADV will be used.

- Accordingly, in the MADV, use all the event properties as a `GROUP BY` option, as that will be used in the subsequent searches on MADV. For example, if a search is going to use the source IP and username as `GROUP BY` options, the MADV should also have the source IP and username as event properties. A MADV can have a lot more event properties than `GROUP BY`.

- Once `GROUP BY` properties are selected for the MADV, they cannot be changed.

- The subsequent search can only filter on properties that are used as `GROUP BY` options in the MADV.

Though creating a MADV is a bit of a complex task, it's highly recommended to extensively use the MADV to optimize the searches. Creating a MADV is like the marketing strategy in food malls, where the most likely picked-up items are placed at the eye level of the customer for the highest visibility.

Summary

In this chapter, we have discussed the crux of QRadar management, which is running searches and maintaining QRadar in such a way that the searches are efficient. We have also discussed various ways in which we can tune the searches. The majority of the issues faced by customers are linked to searches. If you went through this chapter thoroughly, you should now understand the fundamentals of searches, such as the services involved, filters used, indexing, and so on. This should immensely help SOC admins to design searches and SOC analysts to run them.

The concept of SIEM revolves around gathering relevant security information, analyzing the data, and generating alerts for the SOC team. The security alerts in QRadar are called offenses because the alerts are generated when the rules laid down by the QRadar admin are offended. We will discuss the various aspects of rules and offenses in the next chapter.

7
QRadar Rules and Offenses

The greatest challenge for any security team across organizations is to receive informed alerts and then perform incident management. Now, what do we mean by informed alerts? In QRadar, we have discussed how data is collected (*Chapter 4*). What do we do with this data? We correlate this data against the rules that are defined in QRadar.

Rules are security conditions against which every event is matched. If the event matches the rule, the event is tagged with the rule name. If the rule conditions are matched, then an alert is generated. In QRadar, we call security alerts **offenses**. For every offense triggered, we correlate events and flows to break down and explain the offense. So, when it comes to offense analysis, the **Security Operations Center (SOC)** analyst wants to get relevant information about the offense or attack. Once the analyst has the information, the analyst can look up whether it has happened before. If it has happened before, there will be a workflow defined to mitigate the threat.

For every organization, there are some looming threats, and to prevent or mitigate those threats, security use cases are defined. For example, for a financial institution, the security use case might be different than for an organization involved in language training. To define these security use cases, we use rules and building blocks in IBM QRadar.

We will cover the following topics in this chapter:

- Different types of QRadar rules
- Understanding the Rule Wizard
- Historical rule correlation
- Offense generation and management

Different types of QRadar rules

Rules are lists of security conditions that are defined. QRadar has hundreds of rules out of the box. These rules cater to different use cases for cyber-attacks. The use cases are derived from continuous cyber-attacks that happen around the world. The QRadar development team works on different security use cases and continues to add new rules. These rules are added to QRadar via auto updates (we will discuss auto updates later in this chapter). Along with the default rules in QRadar, QRadar users can also create new custom rules based on events, flows, common rules (both events and flows), and other offenses. These are the different types of rules in QRadar:

- **Event rules**: Rules are categorized based on what kind of data is required in the rule conditions. When events are evaluated against certain values, those rules are called event rules. These event rules will consist of different conditions for one or more events that happen in real time.

- **Flow rules**: When rules have conditions based on one or more flows, they are called flow rules. Both event rules and flow rules are capable of triggering offenses if the conditions are met.

- **Common rules**: It may be that we would like to detect an attack based on both events and flows. These are called common rules. As we know about flows, they add more value to the information that we can fetch from events. Events may have limited data based on how the end device is configured to collect the log data. Flow data will be complementary to event data and will help to create better use cases.

- **Offense rules**: You can configure rules to trigger an offense where the condition depends on the statuses of other offenses. This helps when narrowing down fewer scenarios based on certain parameters. For example, you may add other conditions to offense rules to make them more specific.

The following is a screenshot showing the different types of rules:

Figure 7.1 – Different types of rules

On the QRadar UI, we have the **Offenses** tab, which has a **Rules** section, as seen in the preceding figure. Clicking on it, we can see all the rules available out of the box. If we need to create a new rule, we click on **Actions**, and then select the type of rule that needs to be created. Depending on the use case we want to implement, we can choose between different types of rules.

All the different types of rules that we have covered so far are based on the conditions that we have used. An event rule example could be where we define a series of conditions as follows:

- The username is bob
- The source IP is 7.7.7.7
- The event name is Firewall Deny

So, for an event rule, all three conditions will be checked for each event coming in. However, consider a use case where in an organization, the incoming traffic for a certain web server is high between 9 a.m. and 4 p.m. while the rest of the time, the traffic is very low. Now this is a consistent pattern over the years. If one day, the traffic increases after 7 p.m. and is high throughout the night, it would be very difficult to create an event, flow, or even a common rule for such a condition, so there are special types of rules called **anomaly detection rules** for such scenarios.

We learned in *Chapter 6* about saved searches. Anomaly detection rules use saved searches to create a pattern and then find an anomaly in that pattern. That is the reason why the saved searches feature is such an important feature in QRadar. Now, let us discuss the different types of anomaly detection rules:

- **Anomaly rules**: These rules compare the average value over a short interval of time to the average value over a longer period. For example, if an **Endpoint Detection and Response** (**EDR**) solution has sent too many events per hour for the last 2 hours compared to the number of events per hour over the last 100 days, an anomaly is detected. A saved search is used in case of anomaly rules. So, the saved search in this case would be "events per hour" for each log source or log source type. Based on this search, we can create different anomaly rules based on our use case.

- **Threshold rules**: In these rules, a range is specified. If the value is greater or lower than the range, we consider that the threshold (range) has been breached and the offense is triggered. This typically works well when we need to create use cases for monitoring network bandwidth, the number of detected events, and so on.

> **Important note**
>
> Anomaly rules will deal with averages and percentage differences between shorter time spans compared to longer ones, while threshold rules will have a range defined, beyond which the offense will be triggered.

- **Behavioral rules**: This type of rule deals with outliers. A fixed baseline is calculated based on the data received. There are various parameters, such as season (which is nothing but a time frame), current traffic behavior, current traffic trends, and so on. Behavioral rules are based on baselines, which are generated using different parameters. If the upcoming events show a different behavior, offenses are triggered.

> **Important note**
>
> It is very important to understand how a saved search is designed and how the rule is expected to work. As anomaly detection rules work on percentages, it becomes a bit complex to comprehend the rule conditions. Threshold rules are based on ranges and behavioral rules are based on baselines.

In this section, we understood the different types of rules and how to access these rules. In the next section, we will go through the Rule Wizard, which is the tool with which rules can be created or edited.

Understanding the Rule Wizard

As seen in *Figure 7.1*, there are options to create rules. When you click to create a new rule or click on the already available rules, the Rule Wizard opens. The following is a screenshot of what the Rule Wizard looks like. Notice that the rule has different components, such as rule name, rule definition, rule action, rule response, and so on.

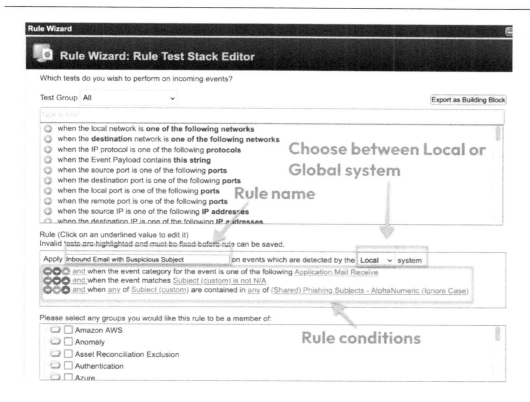

Figure 7.2 – Rule Wizard – part 1

While creating rules, it is important to understand each component and how it can be used. Let's dive in.

Rule name

When creating a new rule, you can mention the name of the rule you want to create. In *Figure 7.2*, the name of the rule is marked as **Inbound Email with Suspicious Subject**. The rule name should actually be a description of what the rule does. It is very similar to writing readable code for any programming language. We use the right kinds of variable names in a programming language so that our code remains readable. Similarly, giving a readable and relatable name to the rule will help you during analysis.

Few customers even have policies on how rules should be named. This helps keep rules segregated and easy to search.

Rule systems

There are two types of systems by which a rule is evaluated. Let us now jump back to *Chapters 1, 2,* and *3* to understand this concept.

Every event processor or flow processor has the `ecs-ep` service, which evaluates incoming events against all the enabled rules in QRadar. Consider an example where there are two event processors in a QRadar deployment. One Event Processor processes events from applications installed on AWS and another event processor from applications installed on the Azure platform. Now, we need a rule for application events where the condition is that if there are more than 20 events labeled **Privilege escalation detected** in 5 minutes, then an offense should be triggered.

As we have 2 processors, QRadar should know whether 20 is the aggregated number for events from both processors or it's 20 events for each processor. To solve this problem, QRadar has two systems in rule evaluation. One is called the **Local** system and the other is called the **Global** system:

- **In the Local system**: The `ecs-ep` service on each processor will look for 20 events in 5 minutes. It does not matter how many events are received on other processors. The `ecs-ep` service just works on the local system and runs the correlation engine.

- **In the Global system**: Once a rule is defined as **Global**, the events that match the rule on each processor are then sent to the console. The console also has an `ecs-ep` service, which then looks for 20 events as an aggregated total from processors. As soon as the console receives these 20 events in a 5-minute window from any or all processors, an offense is triggered.

> **Important note**
> Global rules are expensive in nature as the events are doubly correlated, first at the processor end and then again at the console. As far as possible, restrict the scope of a rule to the local system.

In *Figure 7.2*, we see that the rule has been marked as **Local**.

Rule conditions

This is the main part of the Rule Wizard where the logic of the rule is implemented. This is where conditions are created as per the use case. Different event and flow properties are compared against the values per use case. In *Figure 7.2*, we can see that there are three conditions in this rule. All these conditions are logically ANDed. There is an option to reverse the logic of a condition by using **and not** conditions too.

You can see that there are many options to customize the conditions by using different event/flow properties. You can even use **custom event properties** (**CEPs**) while working on rule conditions. For example, if we have an event where the payload contains an email address but the default event properties do not have a property for an email address, a custom event property called `email address` can be created for the log source. You will need to write a regular expression to parse the email address from the payload. Once done, the email address then can be used in the rule conditions.

> **Important note**
>
> It is recommended that if a CEP is used in a custom rule, the CEP should also be indexed on the **Index Management** page. This helps the rule perform better.

There is another condition called the **timing** condition, which can be used in a rule condition. The **timing** condition is based on the event/flow properties, the number of events/flows, and a certain time limit. For example, see the following:

And when at least 100 events labeled 'File Created' are seen in the last 20 minutes

In the preceding condition, QRadar will keep a tab on the number of events matching the event property, so the **File Created** event name. If there are more than or equal to 100 events with said event name in the last 20 minutes or less, then the condition becomes true.

Another important factor in the rule conditions is the sequence of the rule conditions. The rule conditions, like our search filters (from *Chapter 6*), work like a waterfall model. When the event/flow comes in, it is matched against the first condition. If the event matches the condition, the event is then matched against the second condition, and so on. So, the first condition that we use should be as unique as possible to narrow down the possibility of events matching the rule conditions.

> **Important note**
>
> It is recommended that if you want to use a **timing** condition in a rule, use it as the last condition in the rule condition sequence. This is a way to optimize the rule conditions. The rule conditions are evaluated from the first condition to the last condition and this is also known as the parsing order of the conditions.

Rule actions

Once you have defined the conditions for a rule, the next step is to customize the actions that need to be taken once the conditions are met. When the conditions meet, the offense is generated. In *Figure 7.3*, we see different options when the offense is generated. You can assign a specific severity, credibility, and relevance to the event if the events match the rule. For example, usually, if the events match the condition, we increase their severity. You can have a uniform policy throughout your organization on how to recalibrate the severity, credibility, and relevance of the events based on the kind of rule. This is customizable and should be based on the use case.

Let us discuss an example of an offense. We see data exfiltration from the source IP within our organization. An offense has been triggered for it. From the same source IP, we also see someone has remotely logged in to the web server and privilege escalation is happening. Multiple SOC analysts in a SOC work on the incidents and most of them work independently. In QRadar, security incidents are nothing but offenses. So, in our SOC, two offenses could be triggered, and then be analyzed by two different SOC analysts. This would result in incompetent, inefficient, and incomplete incident analysis.

To help SOC analysts look at the bigger picture of what exactly is happening, we can index the offenses. For this to happen, we must check the **Ensure detected event is part of an offense** checkbox. Once done, you can index the offense based on any event property.

In the example we discussed earlier, it would be wise to index offenses based on the source IP. All the events from the source IP will be attached to this offense. This also helps to chain the different offenses together. So, a few events that might not be significant on their own become a very significant clue for discovering an attack. In our example, the privilege escalation attack and the data filtration attack are correlated. This is like joining the dots to get the bigger picture of the issue. In our case, if we look at the bigger picture, it could be a disgruntled employee who is trying to access confidential data and then trying to send this data over the internet to some other location.

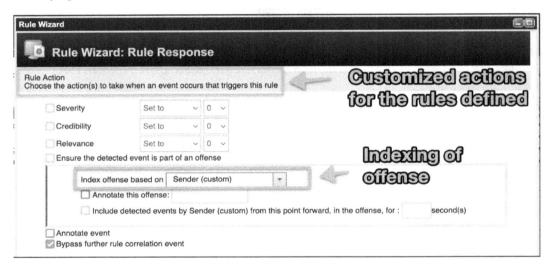

Figure 7.3 – Rule Action

Annotating an offense is nothing more than writing a note about that particular offense. Rules should be defined in a way that avoids a scenario where too many events are attached to an offense; we have the option to configure and include detected events from a particular property for a particular time. This helps us limit the number of events attached to an offense.

The last option we see in *Figure 7.3* is **Bypass further rule correlation event**. What this means is that that particular event will not be matched against any other further rules.

Rule responses

You might be wondering why there are two options, one being a rule action and the other being a rule response, which sound very similar. The main difference between rule actions and rule responses is that whenever an offense is triggered and when the events and the upcoming events match that particular offense, then a rule action is triggered, but a rule response only occurs when the offense is triggered for the first time. The rule response can also be limited using a **Response Limiter** option.

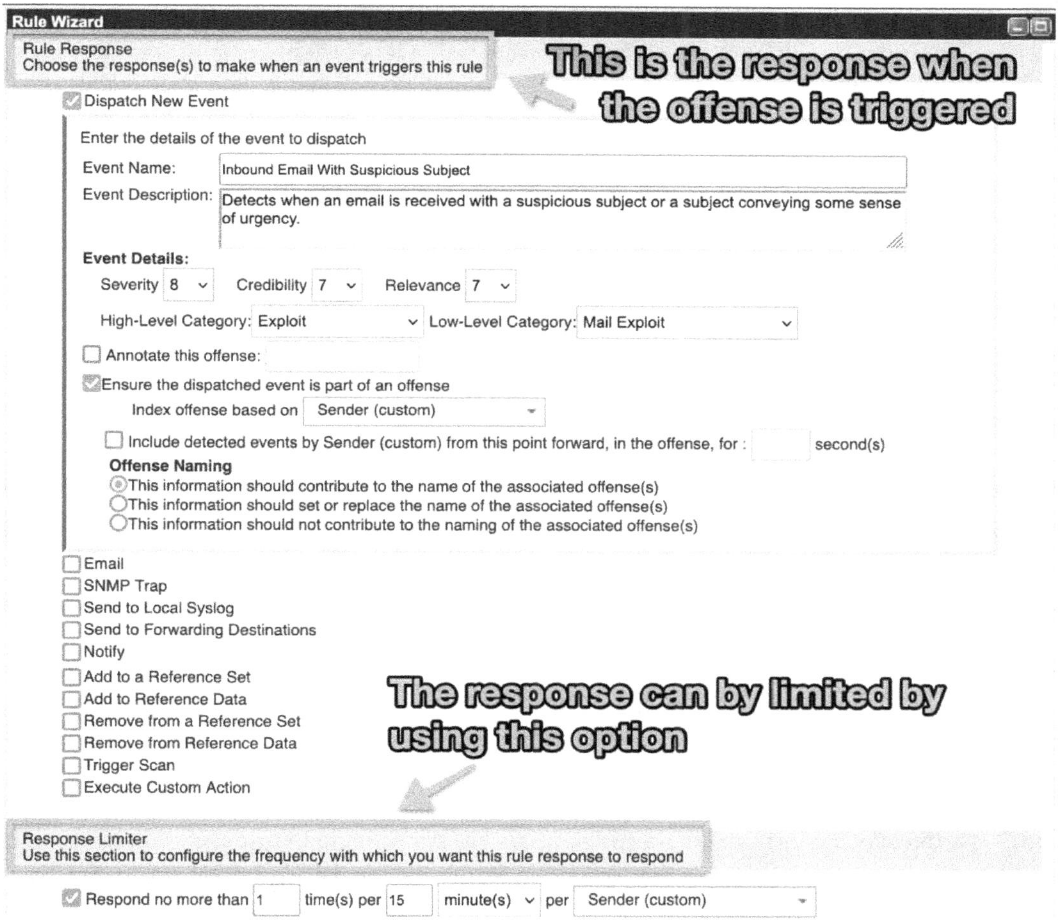

Figure 7.4 – Rule Action and Rule Response

Now, let us look at the options in **Rule Response**. The first option is **Dispatch New Event**. When the rule conditions are matched and the offense is triggered, if this option is selected, then a new event is triggered by the **custom rule engine** (**CRE**). This new event is called a CRE event. A CRE event is a special kind of event. We can give a particular name to this event, for example, in this case, we have named it Inbound Email with Suspicious Subject. You can even assign special **Severity**, **Credibility**, and **Relevance** levels to this particular event. You may also define the **High-Level Category** and **Low-Level Category** settings for that particular event. You can (and we recommend that you do) ensure that this CRE event is part of the offense that has been triggered.

In the preceding example, you can see that the offense is again indexed based on the event property. It is recommended that you use the same indexing parameter that you used in **Rule Action**. This helps keep features such as offense-renaming in order.

One of the most important functions of the CRE event is that it is used by customers for offense-renaming. Usually, an offense is named after the last event that has matched the particular rule – consider that there are hundreds of events that matched a particular rule condition, and the last event that matches the condition name is `Firewall Deny`. Now, the name of the offense will be `Firewall Deny` because it is the name of the last event that matched the particular rule and that triggered the offense. However, this is not a logical solution because the name does not depict the exact attack or the security use case that we wanted to implement. For this reason, there is a feature in QRadar called offense-renaming. What it does is allows us to rename an offense based on the name that we wish to add or that is suitable for that particular use case. In our example, we have used the dispatched event to rename a particular offense. Three different options can be configured to rename an offense. They are as follows:

- When the CRE event name contributes to the name of the offense
- When the CRE event name replaces the offense name
- When the CRE event name does not contribute to the offense name

A detailed explanation of how offense-renaming works can be found here: `https://community.ibm.com/community/user/security/blogs/ashish-kothekar/2021/07/07/how-qradar-offense-renaming-works`.

Rule Response also has other options, such as sending an email, using an SNMP trap, or even triggering a custom script. We can upload custom scripts on QRadar, which can be triggered as a response to a particular offense. This helps us customize a lot of offenses and can help in mitigating those offenses or attacks. For example, suppose we detect that there is a ransomware attack, which might have started in one department of an organization. In such a case, a custom script to immediately take the latest backups from all the systems would be a good option to mitigate the risk of ransomware.

Finally, we see something called the rule limiter or the rule response limiter. This rule response limit can be configured so that you get minimum responses for the particular offenses in the defined timeline.

Using reference data in rules

Before we jump into understanding how reference data can be used in the rules, let us first understand what reference data is. QRadar can integrate with hundreds of different security solutions and use data from these third-party security solutions to create more informed offenses. For example, for a threat intelligence feed can give us a list of IP addresses that are part of a particular **Advanced Persistent Threat** (**APT**) stream. Now, consider that we want to create a rule and we will also want to include a condition as follows: *any of the source IPs match the IP address of these APT streams*. In these scenarios, we store the threat intelligence data in reference sets and then use these reference sets as conditions in rules.

There are different types of ways in which we can store the reference data depending on how it is designed. For example, we have reference maps, reference maps of sets, reference tables, and so on.

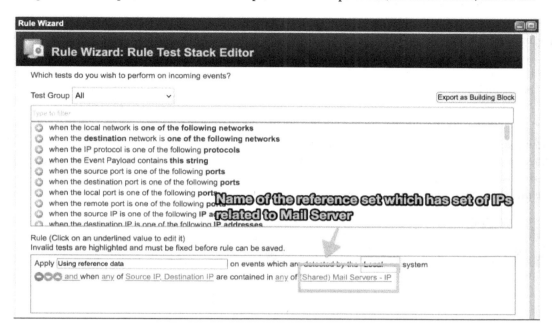

Figure 7.5 – Using reference data in rule conditions

In the preceding figure, we can see a reference set was used. The reference set has all the IP addresses of the mail server. So, if any of the events have either a source IP or destination IP matching the list of IP addresses in the mail server reference set, the condition will be true. Thus, reference data can be used in rules.

Building blocks

Building blocks are rules that do not have any rule response or rule action. So, building blocks are a bunch of rule conditions. Consider a use case that is a bit complex in nature. The easiest way to deal with it is to break the complex conditions into a number of different conditions and use them together. If the simpler conditions have to be used multiple times in different use cases, then it is advisable to convert these conditions into building blocks.

Let us understand this with an example. The conditions for an offense to trigger are as follows:

- The event name is Lateral movement
- The username is any of Disgruntled employees (a reference set)

- The destination IP is any of Crown Jewels (reference set for the IP address of important assets)

- Event from log sources such as firewall

> **Important note**
>
> An event name is associated with a unique ID called a **QID**. A QID is basically a unique number that identifies different events, so QIDs represent a unique event name. Also, every event will have a high-level and low-level category. So, two events with the same event name (but different high-level and low-level categories) will have different QIDs.

Now consider that the first three conditions remain constant and the fourth condition changes for different use cases. The fourth condition could be **Event from log sources like IPs**, **10 similar events in the last 30 mins**, **Source IP geography does not match US**, and so on. In these typical scenarios, the first three conditions could be converted into a building block.

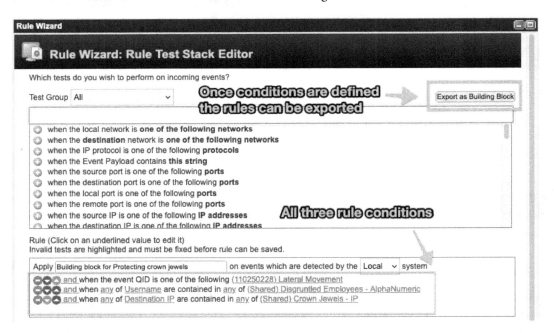

Figure 7.6 – Building blocks

In the preceding figure, we can see all three conditions are defined. To create a building block, click on the **Export as Building Block** button. Once clicked, there is an option to name the building block. An appropriate name should be given to the building block and then it can be reused multiple times in whichever rule is required.

When we talk about building blocks, there is another important feature. Let us first understand a use case and then we will jump to the feature. As a SOC admin, it is important to know that many security incidents are false positives and it may take a lot of time to analyze incidents and then conclude that an incident was a false positive. To avoid this unnecessary strife, QRadar has a feature wherein events and flows with specific QIDs, low-level categories, high-level categories, or traffic directions can be configured to be bypassed for correlation.

When an event or flow is processed, the first rule that it encounters (by design) is as follows:

FalsePositive:False Positive Rules and Building Blocks

What this means is that before the event is run through different enabled rules in QRadar, first, it is checked for being a false positive event. If it matches false positive conditions, then the event is not evaluated further for any of the rules.

The **FalsePositive:False Positive Rules and Building Blocks** rule is made up of a building block called **BB:FalsePositive: All Default False Positive BBs**, as seen in the following screenshot.

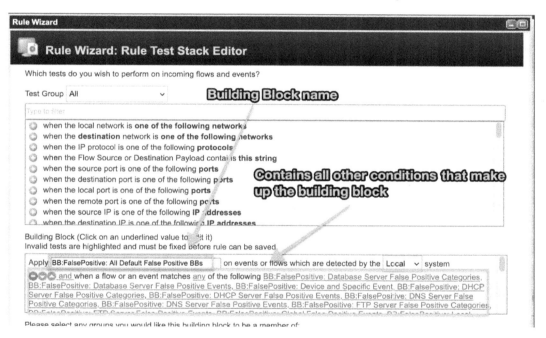

Figure 7.7 – Defining false positive building blocks

We can see the name of the building block and the innumerable number of other building blocks it contains. Building blocks are categorized to contain the deployment-specific information of events and flows, which need not be correlated by the CRE. If the events are correctly marked as false positives, it saves a lot of time and effort for analysts when security incidents are triggered.

What is historical rule correlation?

When you create a rule in QRadar, you would like to test the rule against sample data to understand the number of offenses created, the frequency of the offensees, and the performance impact on the system. The best way to test this is to create a rule and then run a historical correlation to test the rule. The historical correlation rule is run on past data. We can select the range of time for which data was collected to run the rule. For example, a newly created rule can be run on 6-month-old data. Usually, you should run the historical correlation during non-business hours so that the impact, if any, is minimal. Based on the results, you can further tune the rule.

Consider another scenario in which QRadar was down for some time because of, say, a system upgrade activity or any other reason. In such a scenario, we can run a historical correlation and generate the offenses that were lost if there are any.

Historical correlation is also used on your UAT deployments for rule feasibility tests. If too many alerts are generated because of a rule, it would be better to test the new rule on your UAT deployment, tune it there, test it again, and if appropriate, migrate to a production deployment. As such, historical correlation is a kind of retrospective inspection of data against a particular set of rules.

Historical correlation is an important feature in QRadar, as we have discussed. Next, we will discuss managing offenses in QRadar.

Offense generation and management

When the rule conditions match, the rule action triggers. It completely depends on the options selected in the rule action section for an offense to be generated.

> **Important note**
> If all the conditions match and if the indexed property is null, an offense is not generated. For example, if the indexed property is a destination port but the event that triggers the offense does not have a destination port value (the value is null), then the offense is not generated.

This note is for a specific design element with which QRadar ensures that no offense is generated even if rule conditions are matched.

Another peculiar design feature is offense-chaining. Offenses are chained together to provide analysts with a deeper understanding of the security incident. It binds one or more offenses based on the offense index field. This helps analysts relate multiple offenses just by looking at one chained offense. Offense-chaining saves analysts a lot of time spent looking for further related offenses.

Offenses are nothing but security incidents, and every security incident should have a life cycle. There are different phases of an offense depending on the amount of time for which it has been triggered and the events attached to the offense.

An offense remains in an active state as long as events are attached to the offense at less than 30-minute intervals. If the offense is active and no events have been attached for the last 30 minutes, the offense goes into a dormant state. If the offense is in a dormant state for 5 days, the offense becomes inactive. In the inactive state, if new events match the rule conditions, these events are not added to the inactive offense but rather a new offense is generated. After a period of time, defined as the offense retention period, the inactive offenses are automatically removed from the system.

On the other hand, an analyst can mark an active offense as closed once the analyst completes the investigation. These closed offenses are again removed from the system after the defined offense retention period.

> **Important note**
>
> It is recommended that you have as few open/active offenses as possible. This helps your QRadar system perform optimally. As soon as an offense is generated, it can be assigned to an analyst, and once the investigation is completed, the offense can be closed.

In some special cases, an exception can be raised to protect an offense from being closed. A protected offense is not removed from the system automatically when the offense retention kicks in. If the protected offense needs to be removed, then it first must be converted into an unprotected offense and then it can be removed from the system. As suggested previously, do not keep too many protected offenses open, as this will affect system performance.

Every organization deals with security incidents in different ways, but there are a few basics that should be followed:

- Document the offense being triggered.
- Document the security reason why it was triggered.
- Determine whether it was a false positive.
- If it is a false positive, QRadar has a provision to mark it as that, and offenses will not be generated.
- Respond to the security incident. This might be out of the scope of QRadar, or you might even automate this using custom scripts as a rule response.
- Ensure the steps are implemented to avoid or manage similar security incidents in the future.
- Document the response and create a playbook in case similar security incidents are seen again.

When we talk about QRadar, we refer to QRadar SIEM in this book. However, you will also be introduced to another concept – **Security Orchestration, Automation, and Response (SOAR)**. QRadar also has a SOAR product, which is called QRadar SOAR. The previous recommendation bullet points are things that can easily be performed using QRadar SOAR. We will learn about this in detail in *Chapter 14*.

Summary

In this chapter, we dug deep into the fundamental aspects of QRadar, which are rules and offenses. We discussed different types of rules and how these rules can be designed to meet security requirements. We dealt with the minute details of optimizing rules with building blocks and using reference data.

After going through this chapter, you should be in a position to implement security use cases using QRadar rules and offenses. You will be able to use different types of rules, generate alerts in the form of offenses, and manage those offenses.

In the next chapter, we will discuss how internal threats can be mitigated using a QRadar app called User Behavior Analytics.

Part 3: Understanding QRadar Apps, Extensions, and Their Deployment

On the IBM Security App Exchange portal, IBM has published applications and extensions that can be used out of the box. These are like ready-made solutions for the products with which you want to integrate. There are also other apps offering predefined searches, ML integrations, AI integrations, defined rules, building blocks, and so on. This makes it very easy for analysts and CISOs to get a bird's-eye view as well as deep-diving into the technicalities and mining data as required.

There are innumerable apps on offer, of which we will discuss only a few important ones. We will also examine the WinCollect agent and how to manage it. In the final chapter, we will see the fundamental issues regularly encountered on QRadar and how to resolve them. We also provide a helpful list of frequently asked questions in this chapter. We end this book with a small sneak-peek into the new QRadar suite of products.

This part has the following chapters:

- *Chapter 8, The Insider Threat – Detection and Mitigation*
- *Chapter 9, Integrating AI into Threat Management*
- *Chapter 10, Re-Designing User Experience*
- *Chapter 11, WinCollect – the Agent for Windows*
- *Chapter 12, Troubleshooting QRadar*

8
The Insider Threat – Detection and Mitigation

From this chapter onward, we will look in detail at the practical application of what we learned in the last seven chapters. QRadar provides a provision wherein Docker-like applications can be installed, called QRadar apps. These apps vary in nature depending on what type of data they consume and how they use this data to provide value to customers. One such app that we will discuss in detail is **User Behavior Analytics**, also known as **UBA**.

When thinking about securing an organization, we usually think of the threat actors that come into play. Mostly, we think of securing our organization from outside threats by using firewalls, intrusion prevention systems, honeypots, and so on. If we look at the current trends in security breaches, we find that some threat actors are part of the same organization where the breach has happened. These actors are called insider threats.

The UBA app helps us monitor user behavior. If there is any suspicious user activity, the risk score of the user is increased. All this monitoring is assisted by the **Machine Learning (ML)** modules of UBA. If insider threat actors exist, their activity will be monitored, and alerts will be generated if something suspicious is detected. Along with the UBA app, we also have the **Network Threat Analytics (NTA)** app, which monitors the flow activity. NTA can point out anomalous behavior patterns based on the flow of data.

In a nutshell, this chapter will cover the following topics:

- Insider threats – detection and mitigation challenges
- Setting up QRadar UBA
- How does QRadar UBA work?
- Explaining the UBA dashboard

- UBA application tuning

- QRadar's NTA app

- Working knowledge of the QRadar NTA app

Insider threats – detection and mitigation challenges

Detecting an insider threat is like finding a needle in a haystack. The data generated about all the employees working for an organization is enormous. This data could be about the activities they perform and activities that are permitted or not permitted. Also, there could be situations where an activity is suspicious for one employee but legitimate for another. Scanning through all this data to find insider threats is no easy task. But the UBA app helps you meet this challenge. The other two significant challenges that crop up while working on insider threats are as follows:

- **Consolidating multiple identities of the same employee**

    We know that in an organization, we have hundreds of assets that may require a user to create multiple identities. For example, a user, Bob, may have an intranet ID, an ID for accessing Linux servers in a lab, a cloud account for accessing AWS, another cloud account for accessing Azure assets, an account for using applications such as email and the HRMS portal, and so on. Thus, Bob will have multiple usernames that may not look the same because of each application's limitations or requirements. Bob may have the following identities: bob@company.com as the intranet ID, bob_superuser as the Linux cluster username, bob_salesexec as the Salesforce ID, bob_sec_user as the AWS ID, and so on. If Bob is an insider threat, he may be using multiple IDs in parallel to perform a few regular tasks and a few risky tasks. To understand and have a complete picture of what Bob is doing, we need to consolidate all these identities. This helps us create better use cases when we know what different IDs are in use. A few of these IDs could be active, a few dormant, a few suspended, and so on. The normal QRadar rules would not be able to provide such insightful use cases if the IDs are not consolidated.

- **Closely monitoring risky users or some categorized users**

    When users are categorized and monitored, it helps in identifying security policy violations. This is useful to determine and monitor security in terms of segregation of duties and the least-privilege principle. Any organization will have a set of employees it considers risky. These could be disgruntled employees or new employees hired from a competitor organization. All the activities of these users can be closely scrutinized, and any risky action taken by such users may trigger alerts or offenses.

As such, to detect insider threats and mitigate them, IBM has developed the UBA application for QRadar. There are provisions in UBA to consolidate multiple identities. Also, UBA helps in monitoring specific users for a specific amount of time. The monitored users list can always be edited to add or remove users.

What is UBA?

Many security vendors provide similar security solutions, called **User and Entity Behavior Analytics** (**UEBA**). Other vendors, apart from IBM, have the solution as a separate product. But IBM has designed UBA to provide seamless integration with the information already available in terms of event data that we have in QRadar. The integration is tighter than the standalone UEBA solution, providing better results in a shorter period. As QRadar has also introduced App Host as a separate managed host where apps can be installed, the number of computational resources provided to UBA can be drastically increased.

UBA and ML form the backbone of QRadar when it comes to insider threats. These are security products in themselves that are seamlessly integrated into QRadar. There are other vendors that provide UBA as separate products. But adding UBA as an app in QRadar has multiple advantages, including ease of use.

The ML models used in QRadar for UBA help to drastically reduce the false-positive alerts that would have been generated if ML was not used. This helps QRadar analysts save a lot of time when working on incident management.

Setting up QRadar UBA

As we know why we need UBA in our environment, let us look at the steps that need to be taken to set up UBA. The first step would be to install UBA. It is important to know the computational resource requirements of UBA. UBA, when installed with ML, needs a lot more resources than when deployed without ML model, and hence you should always use QRadar App Host.

In the official docs, there's more information on how to install the Machine Learning Analytics app: `https://www.ibm.com/docs/en/qradar-common?topic=app-machine-learning-analytics`

Once installed, we will need to import user information into UBA. Both topics, installing UBA and importing user information, are discussed in detail next.

Installing QRadar UBA

QRadar UBA, as mentioned earlier, is a QRadar app. The app is available from the QRadar App Exchange portal, which is publicly available at `https://exchange.xforce.ibmcloud.com/hub`. The UBA app is available as a ZIP file that can be downloaded for free. Once downloaded, the UBA app can be installed using the QRadar GUI in a very straightforward way. Remember to install the IBM Sense DSM before you go ahead and install UBA. This helps detect the events generated by UBA to be parsed correctly.

As of the time of writing, the UBA app has an integrated ML module. This ML module is used to create behavioral patterns over a considerable time. A time series baseline is created, using which future behavior can be predicted. The baseline can also be used to know what behavior can be considered normal and what can be flagged as an offense. After installing UBA, we would recommend waiting for a day and then installing the **ML analytics app**. This gap provides enough time for the UBA app to create risk profiles before ML algorithms can be run on them.

Importing users into QRadar UBA

After installing UBA, the first thing we need to look at is the addition and consolidation of user identities. For this, we must import users from the following sources:

- **From the LDAP server**: All organizations will have a **Lightweight Directory Access Protocol (LDAP)** server. On a Windows server, the application that maintains this directory is called the **Active Directory (AD)** server.

- **With a CSV file import**: We have discussed how there are various applications in an organization, and for each application, a user might need to create different usernames. All these directories or lists of usernames can be exported in a CSV file. This CSV file can then be imported into the UBA app.

- **Using reference data**: The third way to import users is by using reference data in QRadar. We can store user details in the reference data. Once UBA is installed, we can directly import user data from the reference data.

Except for the CSV file import scenario, the other scenarios, that is, LDAP/AD and reference data import, can be polled. What we mean is, as and when the LDAP directory or reference data is changed over time, these changes will be synced with UBA. Thus, if a user named Alan has joined the organization, over a period of time, his ID from LDAP is added. When a polling interval is hit, the new user ID will be imported automatically to QRadar UBA.

Once users are imported or configurations are set for import, then we should check whether the users can be consolidated. Consolidation is done by tuning the imported users. For each user that is imported, we define a single or multiple attributes that can identify the user. For example, an email address of a user would be the same across all the applications and security solutions. So, email ID is a great attribute based on which usernames can be coalesced. An example of a bad attribute is a department code. A department code would be common across multiple users and hence all users with the same department code would be combined into one username. This is why department name would not be a suitable attribute to consider when coalescing users.

QRadar UBA internals

IBM releases a new version of the QRadar UBA app at regular intervals. QRadar UBA consists of multiple **Custom Event Properties** (**CEPs**), custom rules, dashboards, log sources, reference data collections, and saved searches. This is the beauty of creating and using apps in QRadar. An app can have a customized out-of-the-box solution with so many configurable features in QRadar. Though customized and out-of-the-box are antonyms, we have used them here to signify that the rules, CEPs, and dashboards are customized for the app as well as these customizations being available out-of-the-box. You do not need to spend time creating rules or new CEPs for a particular app or a certain use case to be implemented.

If you want to create more rules for data fetched and generated by the UBA app, you can duplicate the pre-existing UBA rules and then edit them. This provides a starting point rather than having to create rules from scratch.

As discussed earlier, the UBA app also installs the ML app, which is tightly integrated.

You can find more information on ML app installation here: `https://www.ibm.com/docs/en/qradar-common?topic=app-installing-machine-learning-analytics`

With UBA and ML being computationally heavy apps, it is strongly recommended that if you are planning to use UBA, you install App Host and use it to install the UBA app. The UBA app comes with customized dashboards, QID (QRadar Identifier) mappings, rules, log sources, reference data, and saved searches. When running searches related to UBA, it is recommended that you use the saved searches that come with the app. These searches can be duplicated and then modified (as explained in *Chapter 6*). The out-of-the-box searches are optimized, and as a user, you do not need to worry about running the searches in the app.

There is a detailed process view on how the UBA internals work on the page linked as follows. For an end user, the preceding explanation is more than sufficient to kick-start using UBA:

`https://www.ibm.com/docs/en/qradar-common?topic=analytics-process-overview`

How does QRadar UBA work?

As mentioned earlier, before installing the UBA app, there is the prerequisite that a DSM named IBM Sense is installed on QRadar. We learned what DSMs are in *Chapters 4* and *5*. To reiterate, **DSMs** are **device support modules** installed on QRadar, so that QRadar parses incoming data in a consumable format.

When the UBA app is installed, UBA rules are also added. These UBA rules look for certain event data, and if it is found, a sense event is triggered. This sense event is then consumed by the UBA app. The sense event will have a certain risk value as well as a username associated with it. When UBA consumes this sense event, it in turn increases the risk value of the user. Risk scores are stored in the QRadar PostgreSQL database.

The IBM Sense DSM is used to parse the sense event in this case. Once the Sense DSM parses the event, the event is evaluated against all the rules in QRadar, including the just-added UBA rules too. If the event matches the rule conditions as per the configuration, a sense offense can be triggered.

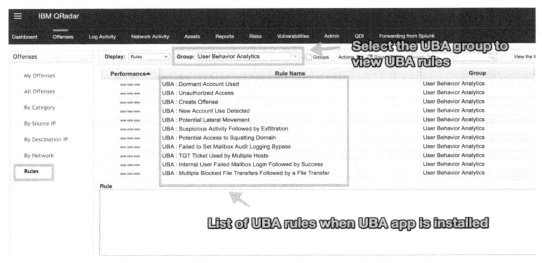

Figure 8.1 – UBA rules

In the preceding figure, we can see the list of the UBA rules added when the UBA application is installed. Similarly, there are building blocks that are also added when the UBA application is installed. The sense offenses that are generated can be seen on the UBA dashboard, which we will look at in the next section. Let's dive in!

Understanding the UBA dashboard

The UBA app displays users and their corresponding risk scores. You can configure the type of users that you would like to monitor. It could be a group of users based on factors such as department, employees on notice period, and employees working for a specific customer across departments. The classification of users is possible when we have enough data on users. This data is usually imported when we integrate with the directory server of the organization.

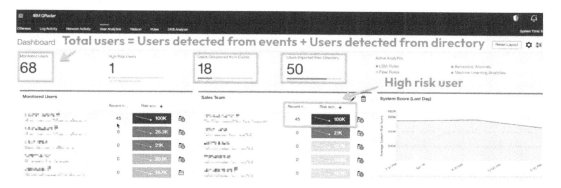

Figure 8.2 – UBA dashboard

In the preceding figure, we see that there are **68** monitored users, out of which **50** users were imported from the directory server. UBA is capable of discovering new users based on the events that it consumes. We previously discussed that the other way to add users is by importing from reference data. In *Figure 8.2*, however, we have not imported any users from reference data.

Along with the users, there are corresponding risk scores associated with them. The risk score of a user is a mathematical number indicating the risk of an insider threat. The higher the risk score, the higher the probability that this user will breach security. Risk scores increase when the monitored user performs dubious or suspicious activity. There are various factors to consider while determining risk scores.

Another important factor to understand is the decay risk. You may have observed that we have discussed how the risk score increases when a risky activity is detected but have never discussed reducing the risk score of users. With decay risk, the risk score of a user decreases if no suspicious activity is carried out by the risky user. The decay risk value is the percentage by which the risk score decreases every hour. By default, the decay risk value is **0.5**. This is a configurable value and you may change it if needed.

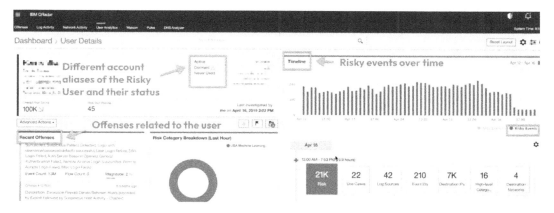

Figure 8.3 – Risky user details

The preceding figure shows risky user details. A single user usually has multiple aliases, which could be either **Active, Dormant**, or **Never Used**. UBA enables us to understand which alias has been used and for what purpose. A risky user's details also show all the recent offenses the user has been involved in. This helps analysts to get a perspective of activities carried out by the user. A timeline is used to show the activities that have increased the risk score of the user.

The activities are further categorized into use cases, log sources, or IPs with which the activity can be identified. By clicking on the categories, even the payload of the event can be analyzed.

The UBA app comes with an ML module; when you download and install UBA, the ML module is also installed as an app. Let us next understand the purpose of this ML app.

Integration with the ML app

The ML app brings with it capabilities of predictive modeling. This application requires intensive computation and works best when you use a separate App Host to host the applications. The ML app is installed after the UBA app is installed.

> **Important note**
> It is recommended that after you install UBA and configure it to import users, you install the ML app after at least 24 hours. This gives UBA enough time to create user profiles and assign risk scores.

The ML app has different models that it uses:

- **Individual (Numeric) user model**: This model calculates a value for a user.
- **Individual (Observable) user model**: This model calculates a set of attributes and their event counts.
- **Peer Group model**: This model is used to build a set of attributes and event counts and alert if the deviation of the user is more for the defined peer group. What we mean by deviation is the deviation in the risk score of the user. This peer group could be all the employees of a particular department. It is expected that the event counts across each group remain proportional. There are multiple ways to define a peer group.
- **Custom model**: A custom model can also be created if none of the previous three models suit your requirements.

Once the ML app is installed and configured, it is time to tune the UBA app. Let us understand the tuning parameters of UBA in the next section.

UBA application tuning

The UBA app, along with the ML app,needs a lot of tuning as per your environment. We have seen that the UBA application has so many configuration parameters. We have already mentioned that if you plan to use UBA, you should install App Host as UBA is a computationally heavy app. The number of resources made available to the UBA and ML apps may limit the number of users that can be monitored. If the number of users becomes high, UBA will require more computational resources, which will in turn hamper performance as the UBA app's graphical interface can become slow or unresponsive.

Some basic tuning tips for the UBA and ML apps are as follows:

- **Import users using a directory server/LDAP/CSV file**

  We have seen that there are many ways in which users are added. In *Figure 8.2*, we saw a few users were discovered using event data such as events and flows. For most of these users, the event data has users such as admin and root that cannot be correlated in QRadar with the actual users. So, it would be wise to only use the UBA application for users that are imported using the directory server. A directory server has detailed information about users, their aliases, departments, email ID, department, and so on. These details should be good enough for UBA to function effectively. In the settings, there is a **Monitor imported users only** option. This ensures UBA doesn't monitor users discovered using event data.

- **Index UBA-specific event properties**

  From previous chapters, we know what search indexing is. When UBA is installed, a few new event properties are also installed. These properties can be seen using the **Index Management** page from the **Admin** tab. You will see properties starting **sensevalue** here. We recommend that you index the following:

  - Sensevalue
  - Username
  - Low-level category
  - High-level category
  - senseOverallscore

- **Understand the Risk Threshold option**

  While configuring UBA, there are two options for selecting the risk threshold:

  - **Static**: The static value is a numeric value above which if the risk score increases, offenses are triggered. When you enable UBA for the first time, the default static value set is 1000. You can reduce the value and observe the number of offenses triggered. If the average risk score of users is around 300 and there are a few users above 600, then you may set the static threshold as 400. These values need to be tuned as per the SOC (security operations center) analysts' findings and thorough research based on your environment. There is no right or wrong value here.

  - **Dynamic**: The higher the value we set for this, the higher the threshold set. The threshold is set based on the activity of all users and the value set. If the value is set high, fewer users will break the threshold and a smaller number of offenses is generated.

As you go on exploring UBA, you should look at the correlation between the option selected for **Risk Threshold**, the value that is set, and the number of offenses that are triggered. You do not want to be overwhelmed by the number of offenses if **Risk Threshold** is not set correctly. Whenever you are tuning **Risk Threshold**, do let the SOC team know that there could be a case of a large number of offenses being generated. Perform this tuning outside of business hours to reduce the risk of losing a legitimate offense in this scenario.

Understanding the QRadar NTA app

We have discussed UBA and understand that it works on different kinds of events that are received in QRadar. But what about flows? Does QRadar use flows to detect anomalies in behavior? Yes, it does. And for that, we have a QRadar app called QRadar NTA. You may install this app from the IBM X-Force App Exchange portal, and it is free of charge.

After installing QRadar NTA, the app trains itself by analyzing the flows already available and creates a baseline of what kind of traffic is received. NTA uses ML algorithms to understand and generate a baseline. The following screenshot shows the configuration settings required for NTA:

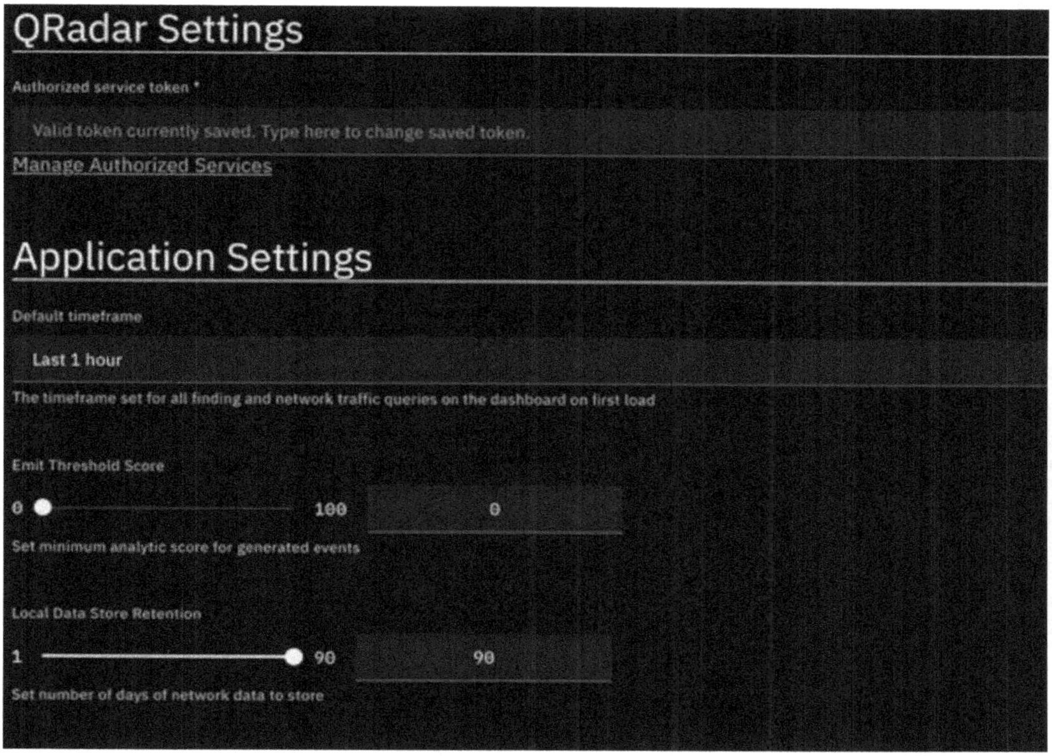

Figure 8.4 – Configuration parameters for the NTA app

In the preceding screenshot, we can see the authorized token that will be generated on QRadar, and then you may copy and paste it into the app.

Default timeframe is the amount of time for which the NTA app analyzes the flow data when it is installed for the first time. If it is set to one hour, it will check the flows from the last hour and generate the baseline. You may increase or decrease it depending on the amount of traffic present in your deployment.

Then, we have **Emit Threshold Score**. Once the baseline is generated, all the consequent flows coming in are measured based on the baseline. If an anomaly is found, events are emitted by NTA for QRadar to know that anomalous behavior has been detected. If the incoming flow has a threshold score of more than what's mentioned in the parameter, an event is emitted by NTA. For example, if the incoming flow is for an adversary action of exfiltrating a large amount of data from an internal database, the threshold score of the flow would be significantly high. Let's say it is 80 for now. If we have set the parameter to 60, NTA will emit an event for this flow. But if the parameter is set to 90, then NTA will not emit an event for this flow anomaly.

The last parameter is **Local Data Store Retention**. It is the number of days for which the NTA app will store the flows. As we know from earlier chapters, flow data can consume a lot of disk space, and hence you should be careful while setting this parameter.

QRadar NTA works on flow records. When similar flow records are found, for example, similar traffic between two hosts, these flow records are grouped into flow sessions. These flow sessions help us to create a baseline. When there is a deviation from this baseline, it is referred to as findings in NTA.

The dashboard of NTA shows information on the flow sessions, baseline coverage, and so on:

Figure 8.5 – NTA dashboard

In the preceding screenshot, we can see the baseline coverage is **99%**, which means that 99% of the traffic was known traffic to NTA and 1% of the traffic was probably from new applications or new hosts.

Along with the basic flow information, the dashboard also shows *findings* data, which is the most critical information in NTA.

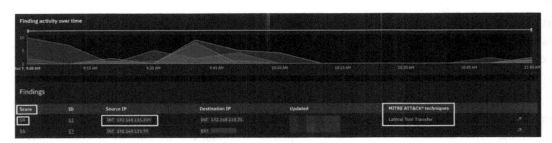

Figure 8.6 – Findings information on the NTA dashboard

For information on the findings, we can see that the flow sessions or flow records are listed along with their score. The higher the score, the higher the deviation from the usual traffic pattern. The flow is also tagged with any MITRE ATT&CK technique if there is one associated with the flow record.

> **Important note**
> The score ranges from 0 to 100.

Double-clicking on the flow record shows further information about the flow. The details page will show information such as flow direction, protocol, total record counts, and bytes. Each flow record is also tagged with deviating categories that would throw more light on what the flow is. This information is essential to determine what exact anomalous behavior was at play.

Another aspect of QRadar NTA is running predefined searches to find anomalous behavior. For example, we can run a predefined search such as *Application is one of the five least common applications by flow count* or *Source or destination IP is one of the top 10 IPs with most bytes transferred*. There are multiple quick filters such as these. These filters help us understand the flows that are out of place and showing different behavior than normal.

QRadar provides us with tools and apps to understand anomalous behavior that may be indicative of insider threats. QRadar UBA and QRadar NTA can be used in tandem to get deeper insights using event and flow data.

Summary

In this chapter, we have discussed how the UBA app can be installed and configured correctly. Always remember that UBA is a heavy application in terms of computational resources and so we should install App Host first. Additionally, the UBA app is updated regularly with new features, new rules are added, and new searches are introduced. To aid with this, always keep your UBA app up to date.

Also, you should use QRadar NTA along with UBA as it helps get granular information if there is anomalous behavior identified. The risk scores provide guidance on what behavior to look at first. The dashboards for both applications will help you detect as well as mitigate insider threats.

In the next chapter, we will dig deep into how QRadar leverages Watson, IBM's cognitive engine, to integrate its AI capabilities with the sea of data that QRadar possesses.

9
Integrating AI into Threat Management

When we look at the challenges surrounding threat management, one that stands out is the prioritization of security alerts. In the process of threat management, we have terabytes of data being analyzed in our local environments and it creates hundreds of security alerts. It is impossible to fully understand the impact of all of these security alerts as well as to prioritize which alerts should be dealt with first in a timely manner. Statistically, it has often taken days to find out a breach has happened, and substantial additional time after that to contain it.

Per IBM's Data Breach Report from 2022, it takes about 277 days on average to identify and stop a data breach. From a security point of view, such a delay in detection and resolution can cause irreversible damage to the organization. So, as the volumes of data and security alerts have increased, so too has the technology become more mature. The introduction of **Artificial Intelligence** (**AI**) has changed the security landscape. The manual interventions that were required to analyze security incidents can now be automated. **QRadar Advisor for Watson** (**QRAW**) is one such solution for this that has revolutionized the SOC analyst's timeline. Where it previously took days and months to analyze security incidents, it now takes minutes or hours.

In this chapter, we will cover three QRadar apps that directly help us to understand the different security use cases and implement them. The QRadar Assistant app is more of a complementary app for the QRadar app ecosystem. QRadar Advisor for Watson and the Use Case Manager app are directly related to how QRadar rules can be tuned and offenses can be prioritized.

To sum that up, we will discuss the following topics in this chapter:

- QRadar Assistant app
- QRadar Advisor for Watson
- QRadar Use Case Manager app

QRadar Assistant app – a quick overview

The QRadar Assistant app gets installed when you install QRadar. This app works like an assistant that suggests and manages all the apps and extensions in QRadar.

The QRadar Assistant app is primarily used to download and install other apps and extensions. For this to happen, if your QRadar appliance is behind a firewall, you will need to access the internet via a proxy. There are different URLs to which you will need to allow access via the firewall. The QRadar Assistant app comes with the Phone Home facility. This app scans through QRadar and will inform you if there are any issues found. No personally identifiable data is collected in this process. By default, you are opted out of this feature. If you choose to opt in, some basic information including offense data, version details, log sources, notifications, and health scores will be shared with IBM.

If there are any privacy concerns, the Assistant app can also be run in offline mode. In offline mode, the Assistant app is only used to manage other applications and extensions.

The QRadar Assistant app also supports multi-tenant environments. But to understand how it does that, you first need to know what a multi-tenant environment is in QRadar and why we might need it.

QRadar can be used by **Managed Security Service Providers** (**MSSPs**) that have multiple clients for which they manage security. So MSSP users will have one or more QRadar deployments and will be ingesting data from multiple clients.

For example, let's say there is an MSSP company called ABC corp. ABC corp is responsible for collecting data from 10 different customers worldwide and providing them with *security as a service*. ABC corp helps customers visualize their ingested data and the alerts generated. But the catch is that each customer should only be able to see their own ingested data and generated alerts, not those for ABC corp's other customers. To accommodate this requirement, QRadar offers multi-tenancy support.

Multi-tenancy can be used by MSSPs and large corporations to segregate data, roles, access, and alerts. QRadar achieves multi-tenancy by creating different domains for different clients. As and when new clients are added, new domains need to be created. But there is more to multi-tenancy than just creating domains. Once domains are created, new users also need to be added, user access has to be configured, new log sources need to be tagged with domains, licensing and license use needs to be monitored, and the rule and network hierarchies need to be managed when QRadar is used for multi-tenancy.

For QRadar apps, different instances of the apps need to be run for different clients. So, ABC corp needs two different UBA instances if they have two clients onboarded. This exact requirement is taken care of by the QRadar Assistant app, which is responsible for creating new instances of the app for different clients and managing them.

QRadar Advisor with Watson

QRAW is based on Watson's cognitive intelligence. IBM Watson uses information from varied sources, breaks it into data points, and then uses the security domain knowledge to stitch together all the data points related to a certain parameter. For example, let's say there is a known hash value for a certain malware, and that hash value is present in the event details in QRadar. IBM Watson uses this information to create graphs and correlations to explain which assets are affected by malware, how this malware entered the organization, how it proliferates or moves laterally, and so on. It also shows the assets and users involved. This kind of detailed analysis is done by QRAW in a matter of minutes. It would have taken SOC analysts weeks or months to get all this information.

QRAW uses Watson's cognitive intelligence. Let us look at how QRAW works with QRadar with the help of the following flow chart:

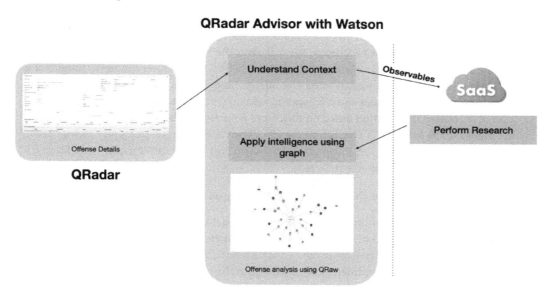

Figure 9.1 – Internal workings of QRAW

In the preceding figure, we see that QRadar collects data and, per the configured rules, creates security alerts known as offenses. These offenses can then be analyzed using QRAW. QRAW runs queries on the QRadar database to gather the required information and create context.

To build context, QRAW requires the following information from QRadar in the right format:

- **CEPs (Custom Extract Properties)**: Ensure that all the CEPs are working as designed and can extract the information correctly from the payload. QRAW is dependent on getting the right information from your CEPs. If there are any issues with CEPs, you should get them fixed.

- **Mapping custom properties**: In a real-world example, there would be hundreds of log sources from which events are received, and these events have resulted in a security incident. These events may have different event properties for each log source defined. There is a need to define and map the event properties to a canonical type. This helps QRAW create context and understand the security incident.

- **Up-to-date network hierarchy**: If the network hierarchy is kept up to date, QRAW is able to understand whether the network is local or remote. This can help generate better graphs for analysts to understand whether lateral movements have been made during a breach or security incident.

- **CRE (Custom Rule Engine) accuracy**: We will cover the MITRE ATT&CK framework when we discuss QRadar Use Case Manager later in this chapter. For now, we can understand that it is a framework with which we can categorize offenses and QRadar rules used. If the CRE rules are accurately tuned, the rules and offenses will fall into their correct categories. QRAW derives a lot of correlated information based on this. Here is the link on how to map the custom rules in QRadar to the MITRE ATT&CK framework in Use Case Manager: `https://www.ibm.com/docs/en/qsip/7.5?topic=qro-mapping-custom-rules-building-blocks-mitre-attck-tactics`

These are the four important factors that need to be tuned before offenses are sent to QRAW for analysis.

So, to understand the context of the offense, QRAW does data mining on the QRadar database and collects the required information as *observables*, which are then sent to QRadar Watson for Cybersecurity, a SaaS solution. These observables are then enriched using a lot of gathered research and this is exactly where cognitive intelligence comes into the picture. These observables are then converted into a knowledge graph and data points. These data points are sent back to QRAW, where the enriched information is then displayed to the SOC analyst. All these steps are carried out within minutes and provide SOC analysts with the right context, insights, and details of the given offense.

There are three types/tiers of SOC analyst:

- **Tier 1 analyst**: Whenever a security incident triggers, it needs to be triaged or categorized. Tier 1 analysts monitor the incoming alerts and if there are no ready playbooks or solutions to the problems, they pass the security incidents on to the next tier.

- **Tier 2 analyst**: Tier 2 analysts work on the security incidents passed to them, analyzing the cause of the issue and the impact of each security incident. They also determine the required action points to avoid these security incidents in future.

- **Tier 3 analyst**: Tier 2 analysts have limited knowledge. Tier 3 analysts are top-of-the-line, highly skilled professionals who work on escalated incidents. They provide guidelines to Tier 2 analysts on how to work on future incidents. Tier 3 analysts also perform threat hunting, which is another part of threat management.

QRAW can help at each of the preceding tiers and stages of analysis.

To add value to what QRAW delivers, it is a great idea to integrate it with a product for security incident management. IBM has a security incident management product called QRadar SOAR. In *Chapter 14*, we will discuss QRadar SOAR at length. For now, let's get a basic understanding of how QRAW can be integrated with QRadar SOAR.

QRAW integration with IBM QRadar SOAR

IBM QRadar SOAR, formerly known as IBM Resilient, is a product for working on incident response. Whenever there is a security incident or data breach, a quick response is required to understand the attack and contain it. Also, it is important to determine a workflow and document the details of the incident. This means that if similar security incidents happen in future, there will already be a workflow defined to mitigate the specific risk. When QRAW is used, the observables from QRAW can directly be used by analysts to create workflows in QRadar SOAR. These workflows can then be used to efficiently manage security incidents.

In short, QRAW feeds observable data to QRadar SOAR to help us build better workflows.

Let us take an example to understand how QRAW can help in a real-life scenario. Let's say an offense is generated after a piece of ransomware is detected in the environment. Now, the offense does not tell us how the ransomware gained entry. It does not tell us which machines are likely to be affected. It does not even tell us if a compromised machine in the environment is acting as a **Command and Control (CnC)** machine or a bot to spread the ransomware.

The CnC machine could be multiple web servers or a single workstation in our scenario. A SOC team with a SIEM system to hand would take days to gather information on a ransomware attack of this magnitude, but with QRAW installed and tuned correctly, all this information will be displayed in a matter of minutes. On the QRAW graph, you will be able to spot machines acting as CnC, the first machine on which the ransomware was detected, and which machine currently has the most connections for that specific ransomware application. On top of that, if you have a SOAR integration, an incident will be created and enriched with all the required information. This also helps generate playbooks for future ransomware attacks.

One question you might have at this point is, *"what is a SOAR playbook?"* A SOAR playbook is a predefined set of steps used by SOC analysts for responding to security incidents. Depending on the type of security incident triggered, a corresponding playbook is made available. So, a playbook for a ransomware attack would be different than one for data exfiltration. With experience, you can create your own playbooks or edit existing ones.

After learning about QRAW, particularly in the context of enriching incidents and improving the analyst experience, let us now dig deep into another QRadar app, called Use Case Manager, which is used for tuning QRadar rules based on the intelligence provided by QRadar.

QRadar Use Case Manager app

While working with QRadar, you might have noticed the enormity of data and the number of rules that a QRadar administrator must work with. The sheer number of rules that come out of the box with the system, and the rules added when new apps and extensions are installed, can be overwhelming. On top of that, some organizations need customized alerts for which custom rules must be created. Managing these rules is a challenge. As we learned in earlier chapters, enabling all the rules is counterproductive and will adversely affect QRadar performance. So how do you manage QRadar rules? The QRadar Use Case Manager app is designed to manage QRadar rules and optimize them. We will look at both these functionalities.

Before we discuss this app in detail, let us understand what the MITRE ATT&CK framework is and how it is used by the QRadar Use Case Manager app.

MITRE ATT&CK is a framework designed for security analysts, threat hunters, red/blue teams, and security execs to help them understand the security posture of an organization. It is a two-dimensional table covering the *tactics and techniques* used by adversaries and hackers.

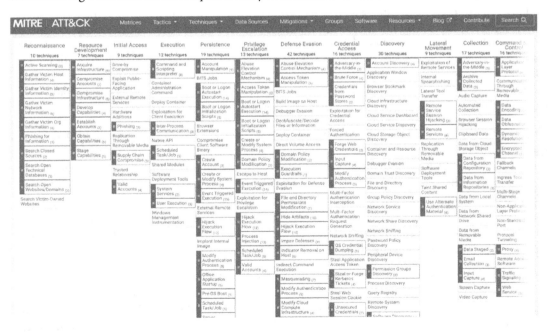

Figure 9.2 – MITRE ATT&CK framework

Figure 9.2 shows a matrix of all important tactics, including **Reconnaissance, Resource Development, Initial Access, Execution**, and **Persistence**. The techniques used for each of these tactics are listed below them. For example, for the **Credential Access** tactic, the first technique mentioned is **Adversary-in-the-Middle**. If you click on the techniques, you will see details on how these techniques are used by adversaries in the wild and how to prevent such attacks on your environments. These tactics and techniques are routinely updated based on new security breaches and incidents as they happen around the world. This framework is open source and is used in many security products. You can explore the MITRE ATT&CK framework further at `https://attack.mitre.org/`.

The QRadar Use Case Manager app also uses the MITRE ATT&CK framework. It maps all the available QRadar rules with tactics and techniques, creating a matrix similar to that of MITRE ATT&CK. Security execs can use this matrix to understand which use cases have been covered using QRadar rules. If the mapping done by the app needs to be corrected, you can do this via configuration.

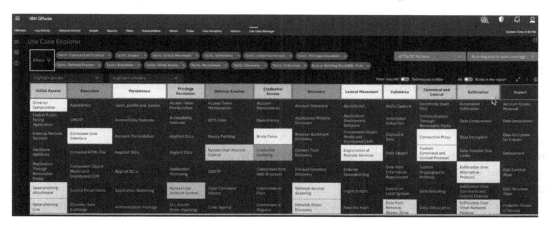

Figure 9.3 – Use Case Manager using the MITRE ATT&CK framework

In the preceding figure, we see the MITRE ATT&CK matrix. By default, QRadar maps a few rules to the tactics and techniques used in the MITRE ATT&CK framework. This mapping can even be edited based on how the rule has been designed. By clicking on techniques, we are shown a list of QRadar rules mapped to the tactic and technique. There are also sub-techniques that can be matched to the QRadar rules.

This provides a bird's-eye view in heatmap form for QRadar admins to understand which use cases, that is, which tactics and techniques, are covered by QRadar.

Another way to use the QRadar Use Case Manager app is for fine-tuning rules. The app can also be used to generate many different types of reports, including the following few important ones to note:

- **Disabled custom properties**: QRadar disables CEPs if they are found to be expensive. Also, as there are usually multiple admins working on QRadar, someone can disable a given CEP used in other rules, reports, or searches. A report of disabled custom properties being used in rules can be generated.

- **Disabled dependencies**: When we design rules, we often use multiple rules or building blocks in one particular rule. If any one of the rules in the parent rule is disabled, the parent rule won't work correctly. The Use Case Manager app helps identify such rules. There could also be building blocks not being used in any rule, which could be deleted.

- **Rule performance**: There are multiple factors such as using regex or payloads, or incorrect sequencing of conditions, that can cause rule performance issues. Using the Use Case Manager app, such rules can be segregated and tuned accordingly.

All in all, the Use Case Manager app helps in editing and tuning rules in a much more sophisticated and efficient way than analyzing rules using the default **Offenses** tab. It also provides reports using which QRadar admins and security execs can get better insights into the complex world of QRadar rules.

Summary

In this chapter, we examined three QRadar apps: namely, the QRadar Assistant app, QRAW, and the QRadar Use Case Manager app. QRadar apps are designed to make it easier to edit, tune, and understand QRadar features. In all, applying artificial intelligence to solve real-life security issues is possible using QRAW. The Use Case Manager app helps you understand and improve the security posture of your organization via a bird's-eye view, along with easy navigation to understand and tune QRadar rules. You should now be capable of using all three apps in your QRadar deployment.

In the next chapter, we will touch on more QRadar applications that are both easy to use and provide a great ROI for the time invested.

10
Re-Designing User Experience

IBM QRadar has been a pioneer when it comes to deep packet inspection and event correlation, providing out-of-the-box rules and hundreds of extensions and apps. It has been the quadrant leader in the Gartner report for the last 12 years. This is no mean feat. But one aspect where IBM QRadar needed improvement was the intuitiveness of the user interface. For the analysts and admins, using QRadar involved a relatively steep learning curve. The dashboards and results had to be analyzed and filtered to make sense. The offense investigations took a lot of effort in terms of the skill and time required from analysts. With all this feedback on the user experience, IBM QRadar came up with the **QRadar Analyst Workflow** app. While working on QRadar Analyst Workflow, a new app specifically for the dashboards called **QRadar Pulse** was also developed and published. The **QRadar Experience Center** app was designed for the end user to simulate attacks and observe what a real-world attack would look like in QRadar. It gives us some perspective and helps customers to build best practices around incident management. It is strongly recommended that if you are using QRadar in any capacity, ensure that you understand these three apps and implement them as directed.

The IBM Exchange portal has hundreds of apps and extensions (some of which are freely available while others require licensing) for all kinds of firewalls, intrusion prevention systems, proxies, DNS servers, and more. But what if someone wants to integrate a log source or create **custom event properties** (**CEPs**) or custom rules around new log sources and analytics, then creating an application would be the way to go. If you have an application based on the **Internet of Things** (**IoT**), it probably won't have an existing DSM in QRadar. For such an application, if you want that IoT device to be covered by QRadar monitoring, the best way would be to create a QRadar application for it. This way if the same application is being used by, say, hundreds of other customers, they would also be able to use your QRadar application (when published on the portal) without much effort.

In this chapter, we will be covering the following topics:

- QRadar Analyst Workflow
- QRadar Pulse Dashboard
- QRadar Experience Center
- Creating your own app

QRadar Analyst Workflow

As we have seen in previous chapters, QRadar is a great security product that handles complex rules, collects different types of event and flow data, provides exceptional results in terms of performance, and helps with regulatory compliance. To enhance the user experience, QRadar Analyst Workflow provides an alternative user interface consisting of the following:

- Search view
- Pulse dashboards
- Offenses view

We will discuss these components in detail in the following sub-sections.

> **Important note**
> IBM QRadar Pulse is an app like any other. It is also part of QRadar Analyst Workflow. This was done as QRadar Analyst Workflow offers a much better user experience than the legacy user interface. Pulse is part of the new user experience.

Exploring the Search view

Analysts need to perform tasks such as event searches, and the **Log Activity** tab in QRadar Analyst Workflow provides multiple ways to run such searches. One of these is via **Ariel Query Language** (**AQL**). AQL is like any other query language where multiple clauses are used in a very specific format. AQL queries can become complicated as more clauses are added, which may lead to errors if the syntax is not well maintained. We can reduce complexity by producing AQL queries as templates that can then be either edited or used elsewhere. For this very purpose there is a feature called **Visual builder** that can be leveraged.

Using Visual builder

When we design and run a search on log activity, we usually add filters including log source name, log source type, source IP address, and so on, using the **Add Filter** option. Every filter added is with an AND condition. An example of this is a search to find events from the last 24 hours from a firewall log source where the username was James. Every condition added to the query reduces the amount of data returned, such as the time period being 24 hours – this restricts the set of logs returned by the query to those from the last 24 hours. The next filter is for firewall logs as the log source, distilling this dataset from the first condition into an even finer dataset. The number of events after each filter either reduces or stays the same, but can never grow. This is because we use the restrictive AND filter while searching log activity. There are ways to include multiple options in filters, such as with the **Equals any of** operator. When we use this operator, we can select multiple values to filter on:

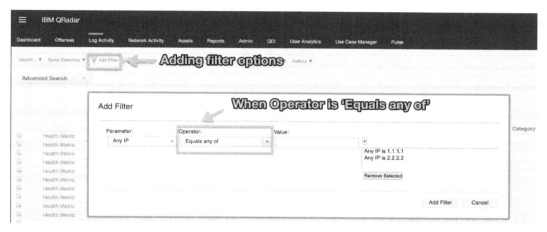

Figure 10.1 – Condition where the values of a search parameter can be ORed

In the preceding figure, we have added a condition where the events returned will match the IP addresses of either **1.1.1.1** or **2.2.2.2**. This is a type of OR condition that can be added to the query with a single filter, allowing different values of the same parameter to be returned.

In IBM QRadar Analyst Workflow's Visual builder, we can logically OR different search conditions. Let us look at an example:

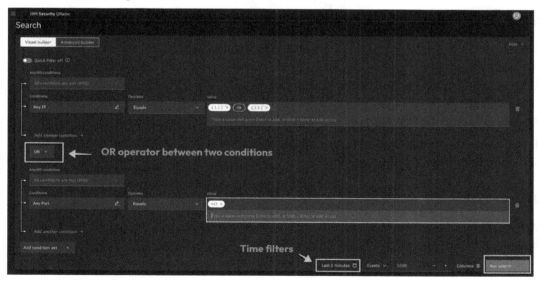

Figure 10.2 – Using the OR operator in QRadar Analyst Workflow

In the preceding figure, we have used two conditions. The first condition is that the IP address should be either **1.1.1.1** or **2.2.2.2**. The second condition is that the port used is **443**.

In QRadar Analyst Workflow, a search can be designed to use an OR operation between different conditions. As mentioned, QRadar Analyst Workflow also has an AND operator that can be used. Thus, a complete filter is as follows:

```
=======================================
Condition 1 - IP address is 1.1.1.1 OR 2.2.2.2
OR
Condition 2 - Port number is 443.
=======================================
```

So, if any of the conditions is true, events will be matched. We also see the time filter set to cover the 5 minutes in *Figure 10.2*, along with the option to specify the maximum number of events fetched by the query.

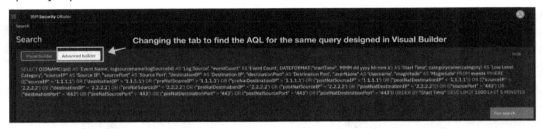

Figure 10.3 – Converting a search into AQL

In the preceding figure, we can see that the query we designed with **Visual builder** can be converted into an AQL query. To do so, just select the **Advanced builder** button to the right of **Visual builder**. The AQL query can then be copied or used for documentation.

This concludes our examination of the Search facility available in QRadar Analyst Workflow. Next, we discuss the **Offenses** view.

Offenses view

We have seen how **Use Case Manager** is best suited for understanding and tuning rules in QRadar. Similarly, the **Offenses view** in QRadar Analyst Workflow is the best way to analyze offenses.

Like in the Use Case Manager app, in the **Offenses** view, we can filter the offenses based on various factors including **Domain**, **Magnitude**, **Severity**, **Status**, **Start Time**, and **Offense Type**:

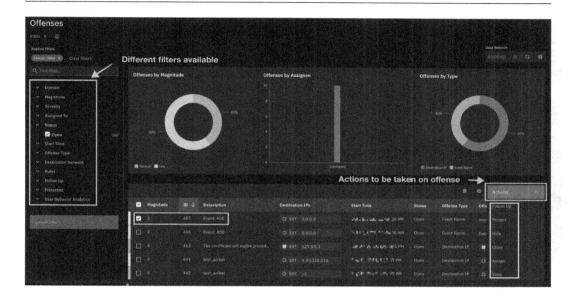

Figure 10.4 – Offenses view

The filtered results are shown on the **Offenses** view dashboard. The graphical representation helps us determine what type of action to take for each offense. This is done by selecting the offense and then clicking the **Actions** button. There are various options available, including **Assign**, **Close**, **Tune**, **Hide**, **Protect**, and **Follow Up**.

> **Note**
>
> If there are too many similar offenses, it is recommended that these offenses are tuned first and then closed later. If there are offenses that you want to prevent from being closed by anyone else managing your QRadar instance, select the offense and click the **Protect** option.

In the next section, we will discuss how to generate custom dashboards with ease and enjoy a data visualization experience far superior to that offered by the previous legacy user interface.

QRadar Pulse dashboard app

QRadar provides a variety of dashboards in its user interface. However, depending on your requirements, these dashboards may need a lot of customization. The problem is that the default dashboards are not very intuitive and often cannot be used as is, while customizing them takes time and research. IBM understood this challenge and created the Pulse app to provide customers with out-of-the-box dashboards for different purposes:

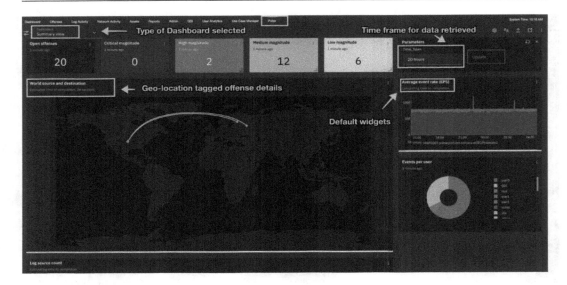

Figure 10.5 – Pulse dashboard app

Important note

QRadar Analyst Workflow has hyperlinks or is connected to other applications including Pulse, User Behavior Analytics, Use Case Manager, and QRadar Assistant. This is done to provide a single pane of glass for all the UI dashboards needed by SOC analysts in their day-to-day work.

In the preceding figure, we can see the **Summary view** of the Pulse application. On this page, we can see the following components:

- **Time_Span**: This determines the time period for which the analyst wants to see data.

- **World source and destination**: In an earlier version of the app, offenses were represented on a rotating globe with their geolocations indicated, showing the source and destination of the offense. For example, a given offense may have a source IP geolocated to one of the islands in the Pacific, and a destination in central Europe. This made it easy to visualize attacks. Now, a two-dimensional map is used to indicate the geolocations of the source and destination IP addresses of offenses.

- **Other widgets**: Other widgets such as average **events per second** (**EPS**), events per user, and log source count can also be customized.

Pulse offers the following dashboards:

- Events and Flow Metrics
- Offense overview
- Offenses
- Summary view

Let us look at the details of the **Offense overview** dashboard:

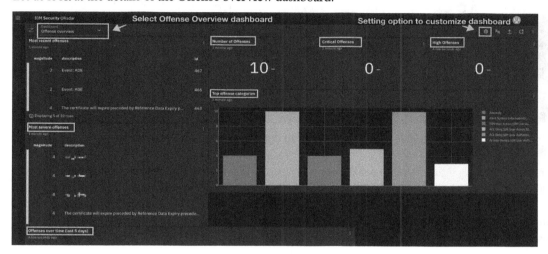

Figure 10.6 – Offense overview dashboard

In the preceding figure, we can see pre-configured widgets, such as **Most recent offenses**, **Most severe offenses**, **Number of Offenses**, **Critical Offenses**, and **Top offense categories**. The **Offense overview** dashboard can be completely customized based on the type of data you want to see. In *Figure 10.6*, we can see the settings button in the top-right corner. Clicking on this button presents us with another window, where we can choose any pre-configured widget or even create a new customized widget. Let us now learn how to create a new widget for our Pulse dashboard:

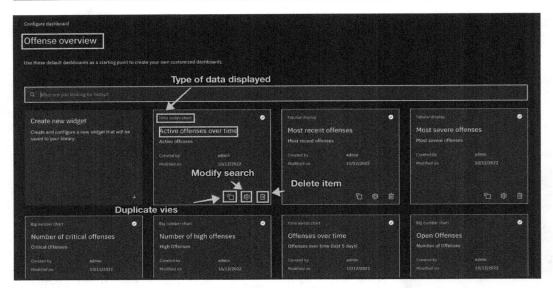

Figure 10.7 – Add a new widget to the Offense overview dashboard

In the preceding figure, we can see different types of widgets available. Different widgets display different types of data. We have highlighted the **Active Offenses over time** widget, for which the type of data displayed is a time series chart. From previous chapters we understand that *time series* means data accumulated over a period of time. There are also options to duplicate the view. This is usually used when we want to make minor changes to the existing view and instead of creating a new widget from scratch. Once the view is duplicated, it can then be edited using the button with a gear icon. The last option is to delete the item if required.

In the next section, we will cover another app that helps QRadar admins simulate attacks and visualize offenses.

QRadar Experience Center

When considering user experience in QRadar, we have to mention the QRadar Experience Center app. This application is developed for customers who are very new to QRadar to help them understand how QRadar works and is best suited for test environments. It provides automated simulations to help us understand how attacks are designed, what exactly happens when attacks occur, and how we can respond.

There are multiple other items imported when you install the QRadar Experience Center app. The current version of the app, there are about 50 **custom event properties** (**CEPs**), 36 custom rules, 16 log sources, and 11 saved searches. All this data is then used by the app to simulate attacks.

After installing QRadar Experience Center, logging in to IBM QRadar, and then going to the **Log Activity** tab, you will see the following screen:

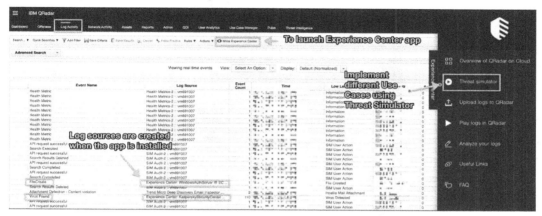

Figure 10.8 – Launch QRadar Experience Center

In the preceding figure, we can see the option to launch QRadar Experience Center on the **Log Activity** tab using the **Show Experience Center** button. Once the app is launched, we see it on the right-hand side of the screen. Also in the same figure, we can see that there are log sources created when the app is installed. The two log sources that we see in the preceding figure are for **WindowsAuthServer** and **Kaspersky**.

On the right-hand side is the **Threat simulator** option, where different use cases are covered. Clicking **Threat simulator** presents you with the following options:

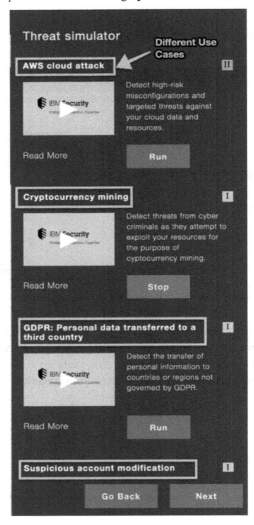

Figure 10.9 – Different use cases covered by Threat simulator

In the preceding figure, we can see the different use cases covered. These range from attacks on cloud services, crypto mining, compliance, and PowerShell to targeted attacks and threats from multiple attacks. For each use case, you can click **Run**, which replays the events and offenses are triggered. These offenses can then be analyzed.

QRadar Experience Center can also be used to upload and run your own logs. For example, you might have custom applications and would like to analyze the security logs from those applications, but there may be no supported **device support module (DSMs)** for your custom application released by IBM. In this case, you can use QRadar Experience Center to upload and analyze the logs. You can also use the DSM Editor if you would like to parse the logs themselves.

QRadar Experience Center provides useful links for different apps and features of QRadar and how to use them. These links provide info on many different aspects of QRadar, including data visualization and analysis, tuning the environment, knowledge bases, forums, and more.

Now that you've learned about the different QRadar apps, let's learn how to build your own app.

Creating your own app

In this chapter, we discussed the different apps available to enhance the QRadar user experience. Even with the level of sophistication offered, there could still be certain ways that you would like to consume QRadar data not offered by these apps. Also, there could be multiple applications for which you require support not yet provided by IBM.

IBM Security App Exchange is the portal on which IBM (among other vendors) publishes applications and extensions that can be downloaded and used by anyone. You need to log in using your IBM ID to access these resources on the exchange portal.

Imagine a scenario where a company named ABC has a software solution for its **customer relationship management (CRM)** needs. The software solution also generates security logs. Company ABC would like to integrate these security logs with QRadar. In such a scenario, company ABC can create a document explaining how to fetch logs from the software solution and send them to QRadar, how to parse new events in QRadar using the DSM Editor, how to write rules, rule responses, rule actions for the events, and so on. This document would be very detailed, and would have to be followed by anyone using the CRM solution and wishing to integrate the security logs with QRadar. These users would have to use the DSM Editor and create CEPs, rules, and so on. This would take considerable time and effort.

An easier solution for company ABC is to create a QRadar app. Once the app is created, it can be sent to IBM for publishing on the IBM Security X-Force Exchange portal. From this portal, anyone using company ABC's CRM solution can download the application and install it in a matter of minutes, with no need to use the DSM Editor nor create any CEPs or rules. This will save a lot of time and effort.

IBM provides a **software development kit (SDK)** for creating apps in QRadar. This SDK can be freely downloaded from the IBM Security X-Force Exchange portal. You can learn more about SDK version 2.0 at `https://www.ibm.com/support/pages/qradar-whats-new-app-framework-sdk-v200`. Anyone can download it and start building their own apps. QRadar also provides an **App Framework** that is updated over time; the latest available version is App Framework version 2. This framework is used when developing your apps. IBM has also provided an example app for use as a reference when creating more complex and advanced apps. The steps for developing a `Hello World` app are described here: `https://www.ibm.com/support/pages/node/6437429`.

It is recommended that you use the template provided in this example app to create your own customized apps, after which you can publish them. This link covers the process of publishing your app/extensions: `https://www.ibm.com/support/pages/publishing-your-extension`.

Summary

We have seen how QRadar is a powerful tool that works with tons of data and different configurations. For the end user, QRadar apps make it easier to visualize and manipulate this data. In this chapter, we have learned how to enhance the QRadar user experience. Creating dashboards, customizing views, and sharing saved searches are a few of the ways to efficiently use QRadar. QRadar Analyst Workflow offers a next-gen user interface to visualize our offense and search data in a completely different way. This new GUI is pretty intuitive too. Similarly, QRadar Pulse provides many pre-defined dashboards. These dashboards can be customized further.

In the next chapter, we will study WinCollect, which is a Windows agent for QRadar.

11

WinCollect – the Agent for Windows

Over the years, there has been no other operating system as popular as the Windows operating system. In the recent past, many IT professionals have adopted Linux over Windows, but still, most machines run on the Windows operating system.

When we think of SIEM and the number of endpoints any SIEM will cater to, we know that the majority of these endpoints will be Microsoft Windows machines. Whether on enterprise servers or desktops, the Windows operating system is popular worldwide.

To cater to this requirement, the IBM QRadar team came up with a Windows-specific agent. This agent is known as **WinCollect**. In this chapter, we will understand the fundamentals of WinCollect, which will be detailed under the following topics:

- Understanding WinCollect
- Types of WinCollect agents
- Tuning WinCollect

Understanding WinCollect

The WinCollect agent can provide centralized log management, highly customized log collection, and security monitoring for all Windows machines. WinCollect can also help us to collect logs from machines by polling logs.

WinCollect can be installed on a Windows machine, and it can even remotely poll events from other Windows machines in a network. These polled events can then be sent to QRadar. Typically, on Windows machines, the types of logs present are application logs, security logs, system logs, custom logs, and so on. It completely depends on the role of the Windows machine. If it is configured as a web server, then there is another category of logs added, called **Internet Information Service (IIS)** logs. So, depending on the services configured and running on a Windows machine, different types of logs can be collected by the WinCollect agent. WinCollect has pre-configured settings to collect Windows data and forward it to QRadar.

The WinCollect agent collects events not only related to the Windows operating system but many Microsoft applications running on Windows machines. Here are a few examples of Microsoft application logs collected:

- Microsoft DHCP log
- Microsoft Exchange Server log
- DNS debug log
- Microsoft IAS log
- Microsoft SQL Server log

Even though all the logs are collected by WinCollect, for parsing, ensure that the latest **Device Support Modules (DSMs)** are updated.

> **Important note**
> There are a few Microsoft applications that do not use WinCollect for collecting data, for example, Microsoft Office 365, Microsoft Windows Defender, and Endpoint Protection.

Windows events can also be collected using protocols such as **Microsoft Remote Procedure Call (MSRPC)**. This is usually done when we do not want to install any agents on Windows machines. But it is recommended to use WinCollect for higher EPS throughput. A very detailed FAQ regarding the use of the MSRPC Protocol can be found here: `https://www.ibm.com/support/pages/qradar-agentless-windows-events-collection-using-msrpc-protocol-msrpc-faq`.

The types of WinCollect agents

Mostly, the WinCollect agent is used for centrally managing event data collection from Windows machines. But, you should know that there are two types of WinCollect agents. One is the widely used **Managed WinCollect** and the other is the **Standalone WinCollect** agent. The basic difference between the managed and standalone WinCollect agents is that managed WinCollect agents can be configured and updated from the QRadar GUI and for standalone agents, the configuration must be done locally on the Windows machine where it is installed. Standalone WinCollect agents come with a Java program that helps to configure agents on Windows machines directly.

Let us understand with examples how managed WinCollect agents work.

Managed WinCollect agents

In the following diagram, we see an implementation of the WinCollect agent in managed mode. We can see that the WinCollect agent is installed on a Windows machine.

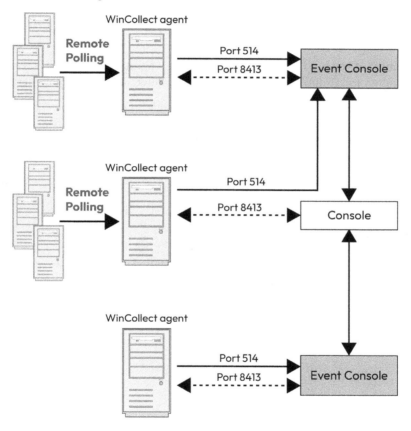

Figure 11.1 – Managed WinCollect agent with remote polling

It can collect data from the local Windows system and can also remotely poll other Windows machines to collect data. WinCollect then sends this data to QRadar, that is, either to an **Event Collector**, **Event Processor**, or the console (it depends on the deployment of QRadar and which components are available).

A managed WinCollect agent can be configured using the QRadar GUI. For QRadar to control the WinCollect agent, command-and-control messages are passed using port 8413. So, port 8413 needs to be open for communication between the managed WinCollect agent and QRadar. When the managed WinCollect agent is installed, we need to mention the configuration server to which the WinCollect agent will communicate for command-and-control. This configuration server can be a QRadar appliance that has the ecs-ec service running. It could be an event collector, event processor, the console, or an event-flow processor.

> **Important note**
> There should not be more than 300 WinCollect agents reporting to one configuration server.

In your environment, if you need more than 300 WinCollect agents (managed), it is recommended that the configuration server duties be divided between the different QRadar components available.

In *Figure 12.1*, we see that WinCollect collects the data and sends the events on port 514, which will be collated by the ecs-ec-ingress service by QRadar. So, when we configure managed WinCollect in our environment, there are two ports that need to be open – 514 for data transfer and 8413 for configuration of data transfer.

Now that we understand managed WinCollect agents, let us understand with an example how standalone WinCollect agents work.

Standalone WinCollect agents

In the following figure, we can see WinCollect agents installed on a Windows machine. One of the WinCollect agents is collecting events locally while the other is collecting events locally as well as remotely.

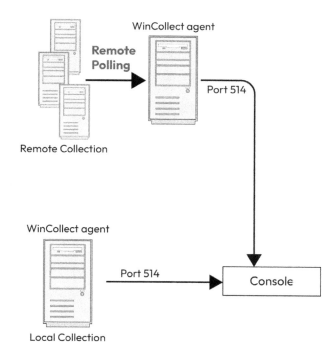

Figure 11.2 – Standalone WinCollect agents with remote polling

In the preceding figure, you will observe that there is no port 8413 for managing the WinCollect agent. The standalone WinCollect agent cannot be configured using the QRadar GUI. The configuration of the standalone WinCollect agent is done using the **WinCollect Configuration Console**.

WinCollect Configuration Console

This is a Java program that is installed on the machine where WinCollect is installed. It has all the options to configure WinCollect, such as adding credentials for other machines, adding destinations, configuring payload size, and so on.

The standalone WinCollect agent also sends events to QRadar on port 514.

In this section, we have covered the two different types of WinCollect agents. We now understand the exact difference between them and have discussed when to use which type of agent. In the next section, we will cover an interesting concept of how tuning can be done for the WinCollect agent.

> **Important note**
>
> WinCollect agent versions are different than the QRadar version, hence before installing WinCollect, ensure compatibility between WinCollect and QRadar versions. There are two versions of WinCollect at the time of writing: WinCollect 7 and WinCollect 10.

Tuning WinCollect

WinCollect comes with many configurable parameters. It has different tuning profiles, polling intervals, and a number of channels. All this is made available to the user to choose the correct option for the amount of data that needs to be collected either from the Windows machine or remotely pulled from other Windows machines.

There are three important parameters for the tuning of WinCollect:

- **Event Rate Tuning Profile**: We know that Windows machines could be our endpoint desktops or could be servers. On servers, there could also be different types of servers. Some could be email servers, web servers, or even DNS servers.

  Depending on the number of events generated per second by a Windows machine, the categorization is as follows:

  - **Windows Endpoint Default**: These are the endpoint desktop machines that produce the lowest number of events per second.

  - **Typical Server**: These are typical servers that generate more events than endpoints. These may include email servers.

  - **High Event Rate Server**: These are servers that generate more events per second than a typical server. These may include web servers.

> **Important note**
>
> The examples that we have provided for Typical Server and High Event Rate Server types are based on events generated per second for the server. If an email server generates more events per second, then it may also be considered a High Event Rate Server.

- **Polling Interval**: WinCollect collects events from Windows machines. It can collect events locally or remotely. The polling interval is the time in milliseconds between WinCollect polling events from a Windows machine.

  The lower the polling interval, the greater the number of times WinCollect will send a query to the Windows machine to fetch events.

- **Number of channels**: There are different types of logs generated by the Windows operating system as well as different applications that are installed on Windows machines. When configuring WinCollect to collect logs from a Windows machine, we need to specify what type of logs we want to collect from a particular machine.

By default, there are three types of events we can see in Event Viewer. These are application logs, security logs, and system logs. Depending on the applications installed, there can also be other types of logs, such as **DNS Server** logs and **Directory Service** logs.

Each type of log that we collect is called a channel. As the number of channels increases, the number of queries made by WinCollect also increases. In *Figure 12.3*, we can see that we have selected two types of logs to be collected. There are two channels that are open for this Windows machine: log source configuration and XPath query configuration.

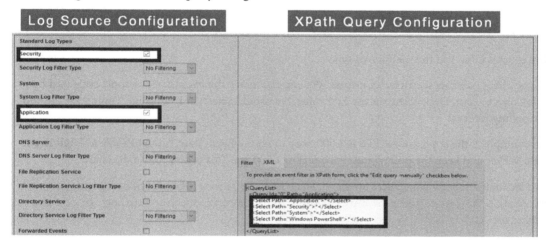

Figure 11.3 – WinCollect configuration: Different channels and an XPath query

Log source confirguation

When it comes to tuning WinCollect, the rule of thumb is that the number of queries made per second should be less than 30.

Now, how do you calculate the number of queries per second? This is given by the following equation:

$$Number\ of\ queries\ per\ second = \frac{Number\ of\ endpoints\ \times Number\ of\ channels}{Polling\ interval}$$

So, for 600 endpoints, if there are two channels open for each endpoint, we will have *600x2 = 1,200* queries. Now, if the polling interval is 10 seconds, then we will have *1,200/10 = 120* queries per second.

Now, 120 queries per second may create performance issues and may throw errors. Our rule of thumb is that we need no more than 30 queries per second. But here, the number of queries per second is 120.

To tune our environment in such a scenario, we can increase the polling interval to 40 seconds. Now, our queries per second will be *1,200/40 = 30* queries, which is what is expected.

So, to tune WinCollect settings, you can modify the number of endpoints, the number of channels via which queries are sent, and the polling interval.

Another important factor when configuring the WinCollect log source is the Event Rate Tuning Profile. We have tested the profiles and polling intervals extensively and have come up with a chart that shows the best combinations that you can choose.

How do you use the chart? There is a chart which you can find on the link:

`https://www.ibm.com/support/pages/wincollect-lets-talk-about-log-source-event-rates-tuning-profiles-updated`

The chart is table 1 in the mentioned link.

In the chart, we can see three columns. We can calculate the events per second collected by a WinCollect agent. Depending on the EPS value, we should select an Event Rate Tuning Profile and the polling interval.

For example, if the approximate EPS is 1,000, we can set the Event Rate Tuning Profile to **High Event Rate Server** and keep the polling interval equal to 1,500 ms. This is how to use the chart.

This technote explains WinCollect tuning in detail: `https://www.ibm.com/support/pages/WinCollect-lets-talk-about-log-source-event-rates-tuning-profiles-updated`.

XPath query configuration

If, for certain Windows machines, we do not wish to collect all the application and the system logs but need a very specific kind of logs, we can use **XPath queries**. In *Figure 12.3*, we can see XML expressions suggesting the kind of logs that we would like to collect. XML can be designed in Windows Event Viewer using the **Create custom view** option.

It is recommended to understand the type of logs required and then design the XPath query accordingly. Detailed steps on how to create XPath queries can be found here: `https://www.ibm.com/docs/en/qradar-common?topic=xpath-creating-custom-view`.

Now that we know what WinCollect agents are and have learned how to tune them, let's summarize the learning outcomes of this chapter.

Summary

WinCollect agents make it easier for QRadar admins to collect required data from Windows machines. Using other protocols such as MSRPC may present certain challenges, which are then addressed using WinCollect agents. Therefore, to collect events from Windows machines, the WinCollect agent is the recommended solution. In this chapter, we have seen different types of WinCollect agents and understood the different scenarios to use them in. In addition, we have dug deep into tuning the WinCollect agent for optimal performance.

In the next chapter, which will be the last chapter of the book, we will cover QRadar troubleshooting, frequently asked questions about QRadar, and the next-generation look of QRadar.

12
Troubleshooting QRadar

In the previous chapters, we discussed the architecture of QRadar and walked through how to use QRadar and its various features. In this chapter, we will discuss the common problems or issues that you may face while working on QRadar. QRadar has evolved a lot over the last decade. There have been regular updates to the underlying **operating system** (**OS**), new features have been introduced, and bugs have been resolved. Also, all the vulnerabilities found in the product are addressed in the update packs and version upgrades. Over the years, common issues were discovered that happened mostly because of the complexity of the product and a lack of understanding of the configuration details. The common problems can be categorized as follows:

- Log source and flow integration issues
- QRadar deployment issues
- QRadar app issues
- Performance issues

Over the years, I have found that QRadar admins struggle with a few basic queries. That could be because of a lack of information available or a lack of experience with the product. For QRadar admins, I have curated a **frequently asked questions** (**FAQs**) list with answers and related links. Hopefully, you will find this information useful.

In the last section of this chapter, we will look at how QRadar is likely to change over time. We will discuss a new platform where QRadar could be used as a **security information and event management** (**SIEM**), along with other security products.

We will cover the following topics in this chapter:

- Exploring log source and flow integration issues
- QRadar FAQs answered
- A next-generation QRadar sneak peek

Let us dig deeper into the common issues seen in QRadar.

Exploring log source and flow integration issues

As we know, QRadar ingests data that in the form of events and flows. As such, let's look at issues related to log source and flow integration in QRadar. We will begin by discussing the autoupdate issues you might face and then move on to log source configuration issues. You will also be provided with resources such as the QRadar DSM guide and the IBM QRadar Community forum for troubleshooting purposes. Finally, we will cover flow integration issues, explaining various configuration parameters related to flows and providing resources to understand and customize flow parameters. Let's get started!

Autoupdate issues

For the log source integration, we know that QRadar uses different protocols and DSMs. QRadar's **autoupdate** feature is responsible for updating these protocols and DSMs, provided it is configured.

Autoupdate is a feature wherein QRadar reaches out to external IBM servers to download the latest updates for log source DSMs, protocols, QID mappings, internal scripts, and so on. However, there could be network issues if your version of QRadar is behind a firewall. If your organization is using a proxy, ensure that proxy details are correctly filled in. If your proxy has **secure socket layer** (**SSL**) inspection enabled, QRadar may fail to connect to the Autoupdate server.

One of the resources to understand and troubleshoot Autoupdate issues is available at `https://www.ibm.com/support/pages/qradar-important-auto-update-server-changes-administrators`.

Log source configuration issues

IBM QRadar provides a QRadar DSM guide to help customers configure log sources for event data. All the log sources that are mentioned in DSM guide are fully supported by the IBM QRadar Support team. So, if there are any issues while integrating the log sources mentioned in the guide, you can go ahead and open a case with the IBM QRadar Support team.

For log sources that are not mentioned in the DSM guide but that you would still like to integrate, you can use the DSM editor to create a parsing log and map the events to your customized log source. While performing this activity, if you have any issues, then you should discuss this in the IBM QRadar Community forum. IBM support cases are exclusively for log sources that are supported – that is, those mentioned in the DSM guide.

In this context, some important resources are as follows:

- **DSM guide**: The guide can be found here: `https://www.ibm.com/docs/en/dsm?topic=configuration-qradar-supported-dsms`. The DSM guide is updated and published every month, so new log sources may be included in the updated version. Always use the most recent DSM guide when referring to it for log source integration.

- **IBM QRadar Community forum**: The forum can be found here: `https://community.ibm.com/community/user/security/communities/community-home/digestviewer?communitykey=f9ea5420-0984-4345-ba7a-d93b4e2d4864&tab=digestviewer`.

  Apart from discussing your queries on QRadar and integration, you will also find that this community has a **Blogs** section. There are more than 300 blogs that discuss the various features of QRadar as well as log source integrations. This is an excellent resource for meeting different like-minded professionals who may have worked extensively on QRadar.

- **IBM Request for Feature Enhancement**: `https://www.ibm.com/support/pages/qradar-request-enhancements-rfe-and-how-use-them`.

Another aspect that should be discussed is the **Request for Feature Enhancement** (**RFE**), as mentioned in the preceding list. Customers can fill in a form and explain why they need a certain feature or new product DSM integration. Once the form is filled in, it goes to the IBM QRadar product management team, and they can take a decision on whether to include the new feature and, if so, in which future version. The QRadar Community and QRadar RFE are not only limited to log source integration issues but all QRadar features available and requested.

Lately, the RFE process has been optimized and has a new name, **IBM Ideas**. Here is the link to the IBM Ideas Portal: `https://www.ibm.com/support/pages/welcome-ibm-ideas-portal`.

Flow integration issues

In *Chapter 4*, we discussed the event pipeline and described how DSMs and protocols work in detail. When you consider flows, there are many configuration parameters that come into the picture because of features such as deduplication, super flows, flow direction, and so on. It is important to understand these concepts related to flows before we integrate and customize flow parameters. We discussed these concepts in previous chapters, but we have not discussed the different parameters of flows that can be configured. To understand the flow parameters better, you can refer to this resource: `https://www.ibm.com/support/pages/node/5691056#flow_config_parms`.

On the link, you will see multiple parameters defined. Each parameter has a specific functionality. For example, `SV_VERIFY_SEQUENCE_NUMBERS` is the parameter that specifies whether to use sequence number verification to detect when messages are dropped. The value of the parameter would either be `Yes` or `No`, depending on your requirement. The parameters will help you configure your flow ingestion in a customized way.

Many of these parameters can be configured in an `nva.conf` file, which is present on the Console in three different locations. The one that needs to be modified is `/store/configservices/staging/globalconfig/nva.conf`.

Once the changes are made, a full configuration deploy needs to be done.

> **Important note**
> **Full configuration deploy** pushes the deployment configuration to all the managed hosts in an environment.

The other factors to look at in the flow integration issue is the network connectivity between the flow source (the end source from where we are collecting flows) and the flow collector or QRadar network insights. There are tools such as tcpdump that can also be used to verify network connectivity issues.

This leads us to the next section, where we will discuss deployment issues in QRadar.

Understanding QRadar deployment issues

The second most common issue seen in QRadar is related to deployment. Deployment issues range include adding managed hosts, adding high availability hosts, managing high availability, and troubleshooting normal/full deployment issues. The following are the three most important deployment issues we have seen in the field:

- When we talk about deployment, the most important concept to understand is the difference between normal deployment and a full deployment in QRadar. To make it easy, normal deployment changes will only affect the services to which the changes are being made. Not all the services on the managed hosts and Console are restarted because of deployment changes, but when you do a full deployment, a complete set of configurations is pushed, resulting in the restart of all services. Prior to version 7.3.1 of QRadar, the data collection service was `ecs-ec`. Later, we introduced `ecs-ec-ingress`, which remains isolated or unaffected even when a full deploy is run. Other services are affected, but not the data collection service.

  To understand the nuances of the difference between deploy and full deploy in detail, IBM published a IBM tech-note, which is available at `https://www.ibm.com/support/pages/qradar-what-difference-between-deploy-changes-and-deploy-full-configuration`.

- Another concept that we should be familiar with in QRadar is **High Availability (HA)**. Ideally, when we want redundancy from appliance failures, we add HA hosts to the Console or managed hosts. When we add an HA host, data present in the `/store/ariel` is replicated between two hosts using **Distributed Replicated Block Device (DRBD)** technology. The configuration data between the hosts is kept in sync using the **rsync** tool. Once HA is configured, data and configuration info is transferred automatically. No human intervention is required unless HA breaks down.

When two appliances are in HA, one of them can be active mode and the other one in standby mode. Only one appliance is active, and if that active appliance fails, the other appliance takes over and starts required services. There are many other states that appliances or hosts could be in when HA is configured. A very detailed IBM tech-note on how to configure and troubleshoot HA in QRadar can be found here: `https://www.ibm.com/support/pages/node/6565347`.

When HA fails, you can always go ahead and open a case with IBM Support. However, before that, you can troubleshoot HA using a script provided by QRadar. The script is called `ha_diagnosis`. This script can be found on your QRadar appliance at `/opt/qradar/ha/bin/ha_diagnosis`. A IBM tech-note describing how to use and troubleshoot HA issues is available at `https://www.ibm.com/support/pages/node/6828547`.

- The third issue is related to adding, editing, and removing managed hosts in a QRadar deployment. When a managed host is added, the adding process can fail because of various reasons. Again, we should understand the configuration parameters available and how deployment works. In earlier versions, when a managed host was added, you had to specify whether the connection between the managed host and Console was encrypted or not. However, in QRadar 7.5 version and above, by default, the connection is encrypted between the managed host and Console.

When we discuss deployment, it is good to know which ports are required by QRadar components to interact with each other. Here is a list of ports that need to be opened for different QRadar components: `https://www.ibm.com/docs/en/qradar-on-cloud?topic=qradar-port-usage`. If the ports are not opened, the connection will break, and there will be issues when adding managed hosts, the status of the managed hosts could change to unknown, and so on. The list of port usage is very specific and detailed. Go through the linked documentation before deploying QRadar, and keep the link handy if there are issues in deployment.

> **Important notes**
> After adding or removing a managed host, it is a good practice to run a full deploy on QRadar.

If you want to check whether there are any outstanding issues in your deployment, you can validate them by using an internal script. The script is available at `/opt/qradar/support/validate_deployment.sh`. You can run the script and check the output to see whether there are any bad entries. It is always advisable to open a support case with IBM Support if there are any deployment issues seen in the script output. Also, there are many IBM tech-notes related to deployment failure, such as the one available at `https://www.ibm.com/community/qradar/home/deploys/`.

In the preceding URL, we can see eight steps mentioned in the checklist, each with a reason why a deployment may fail. Go to the link and then click on each point to go into further details on each point. Performing the changes mentioned in the IBM tech-note is safe and are expected to be performed by QRadar administrators. Also note that there are other IBM tech-notes mentioned in the link, explaining the defects associated with deployment issues, common questions, and common error messages.

With that, we have covered the difference between deployment and full configuration deployment, and the various issues commonly faced by customers while deploying changes in QRadar. In the next section, we will understand commonly seen issues seen when working with different QRadar apps.

Investigating QRadar app issues

The IBM X-Force Exchange portal hosts multiple apps and extensions that can be installed on QRadar. QRadar has an application framework that is utilized when apps are deployed. There are many services involved in application framework functionality. The following IBM tech-note details each service involved in application framework along with the ports they use: `https://www.ibm.com/support/pages/node/6190995`.

If there are issues installing applications or any other issues with applications, it is a good practice to check the status of these services. When we talk about QRadar applications, we should also discuss **QRadar App Host**. As we know, App Host is added to a deployment so that the Console can offload the work of managing and hosting apps to App Host. The Console can then work exclusively on QRadar functionalities. If you are using QRadar apps such as **User Behavior Analytics** (**UBA**) with machine learning or QRadar Watson Advisor, it is recommended that App Host is deployed in an environment. These apps are computationally heavy and work best when they have ample resources available.

The IBM tech-note explains the different micro services for an application framework in QRadar: `https://www.ibm.com/support/pages/qradar-services-responsible-applications-and-application-framework-functionality`

In the preceding URL, we can see the services, their description, where these services run, the port number utilized, as well as the command that can be used to check the status of each service.

When we talk about applications, we need to mention the containers that are spawned from the images that have been downloaded. If something is not working as expected in an application, it is recommended to check the container logs for that particular application. A `journalctl` command can be used along with the container name to view the logs related to the application. Check out this IBM tech-note to see how to view logs related to an application: `https://www.ibm.com/support/pages/node/6208418`.

Exploring QRadar performance issues

As discussed in *Chapter 4*, an event pipeline is the number of services and functions that an event passes through. So, consider an event pipeline as a highway and the services and functions involved as the speed breakers or hurdles on the highway.

The following are four conceptual hurdles that can occur in an event pipeline:

- The first hurdle that the event has to cross is reaching the highway – that is, reaching the event pipeline. Events are ingested into QRadar using different protocols. Each protocol has a limited number of events that can ingested per unit time. This is also called a **protocol queue**.

  Consider that, on an Event Processor, there are events being collected just for one protocol, such as an eStreamer protocol for CISCO. Even though the total number of events per second is much fewer than what is permitted by the license on the Event Processor, there could still be a performance issue on the Event Processor. In such scenarios, contact the IBM Support team, and they will help you increase the queue size for the protocol.

- The second hurdle is eps being more than the licensed value. If the number of events per second is more, QRadar pushes the events into an event buffer and waits for the event spike to subside. If the event spike subsides, the events are pulled from the event buffer and then processed. However, if the event spike does not decrease, then QRadar may drop the events and send a notification to the user on the QRadar user interface. At this stage, we should increase the license, considering the specifications of the appliance, or add a new appliance and then allocate the required license from the license pool.

- The third hurdle could be parsing events. If the events are not in the correct format as expected, or the DSM that we are using to parse the events is not the latest version, then it may take time for events to parse, or they might even go unparsed. In such a scenario, performance degradation may happen. Another factor related to parsing is **custom event properties** (CEPs) that are extracted from the event payload. If the CEPs are not defined optimally, they can also lead to performance issues. At the time of the performance degradation scenario, there are a couple of scripts that can be run to collect the diagnostic data, which can later be sent to IBM Support for review. The two scripts are as follows:

  - `findExpensiveCustomProperties.sh`
  - `dumpDSMInfo.sh`

  They are both located in the `/opt/qradar/support` directory. Both scripts create an output file that should be shared with IBM support.

- The fourth hurdle is rules and building blocks and their correlation with each event. Every processor and Console will correlate the events received. If the rules are not optimally designed, then we will receive a performance degradation notification. This notification will also have a list of rules and building blocks that might be problematic. QRadar provides a visual performance stat in the **Offenses** tab too:

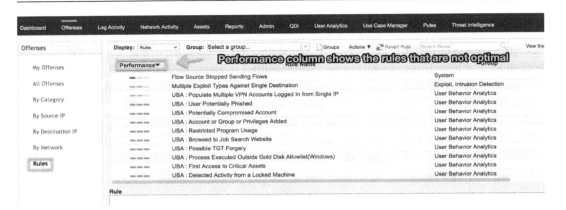

Figure 12.1 – The Performance column for the rules that are not optimal

In the preceding figure, we can see that the **Performance** column shows three bars. It is either red (a single bar highlighted), orange (two bars highlighted), or green (all three bars highlighted). When the bars are red and orange, it indicates that the rules had performance issues. Now, a rule can have performance issues because of many reasons, including the following:

- The rule uses reference sets, and the number of elements in the reference sets is more than 50,000. If the number of reference set elements is high, then it takes a longer time for the rule to correlate the conditions, thus increasing the chances of performance degradation.

- The rule has a timing condition, but this is not the last condition in the rule.

- When we use a timing condition in a rule – for example, 100 events in 5 minutes – the rule becomes a stateful rule. For every event received in a stateful rule where the timing condition is the not the last condition, QRadar will need to maintain the state of the rule for other conditions too. This adds up computational resources for rule evaluation.

Apart from rules, other factors to consider for performance tuning are as follows:

- Make use of the **Use Case Manager** to tune QRadar rules. Automated suggestions are made by the app on different factors.

- Limit the number of open offenses. Too many open offenses lead to performance issues.

- Tune building blocks by maintaining authorized servers, discovering servers, and updating the network hierarchy. This will help to provide more efficient results.

IBM QRadar provides a host of different tools and scripts that can be used to troubleshoot QRadar. This link provides a list of scripts and tools that can be used: https://www.ibm.com/community/qradar/home/tools/. The most notable among them is the get_logs.sh utility. This utility can be invoked from the command line or even used from a user interface to collect QRadar logs.

We have touched upon many aspects of troubleshooting QRadar. On many points, we have suggested opening a case with IBM Support too. Use this IBM tech-note as a reference point when opening a case with IBM Support: `https://www.ibm.com/support/pages/getting-help-what-information-should-be-submitted-qradar-service-request`.

Till now, we have discussed troubleshooting techniques in detail. In the next section, we will look at different QRadar queries that customers have and how to solve those real-world problems.

QRadar FAQs answered

We have come a long way and covered all fundamental aspects of IBM QRadar. In this appendix, we will cover all the major queries that QRadar admins and SOC analysts will have when working with QRadar.

Query 1

What are the other major IBM security products that can be integrated with IBM QRadar?

In this book, when we talk about IBM QRadar, we also mean the IBM QRadar **Security information and event management (SIEM)** solution. QRadar SIEM deals with collecting data in terms of flows and events and generating security alerts. Out of all the IBM security products, IBM QRadar SOAR best complements IBM QRadar SIEM.

SOAR stands for **security orchestration and response**. IBM QRadar SOAR is a different product (from QRadar SIEM), which was previously known as Resilient. The offenses that are generated in IBM QRadar SIEM are sent to IBM QRadar SOAR, where they are known as incidents.

QRadar SOAR can integrate with different products too, to receive security incidents. For example, if an analyst observes some anomaly, a simple email from the analyst on a pre-specified email ID can generate a security incident on QRadar SOAR. QRadar SOAR manages the complete incident life cycle from the time it was generated, enriched by data from different sources, and automatically assigned to an SOC analyst to it being responded to and then documented.

QRadar SOAR has a feature to help dynamically code the response to an incident. Different workflows can automate the process of the response, thus saving a lot of time on repetitive tasks. The responses include informing stakeholders by email or on Slack, blocking IP addresses and URLs, and changing configurations on proxy.

IBM QRadar SIEM and SOAR complement each other and provide great value to customers.

More details on SOAR can be found here: `https://www.ibm.com/products/qradar-soar`.

Apart from QRadar SOAR, QRadar SIEM can be tightly integrated with IBM ReaQta, which is IBM's **Endpoint Detection and Response (EDR)** solution.

Query 2

Are there other SOAR products that QRadar can integrate with?

Yes. IBM QRadar integrates seamlessly with other SOAR solutions such as the following:

- Demisto (now known as Cortex XSOAR)
- ServiceNow
- Splunk SOAR

Query 3

Can multiple IBM QRadar deployments be integrated together?

Yes. There are instances where a single customer may need to install multiple QRadar deployments because of compliance or other network restrictions. In such scenarios, we can use QRadar Master Console.

QRadar Master Console can pull data from multiple QRadar deployments and show it on a single pane of glass. We can fetch offenses from different deployments, the managed hosts' status, CPU utilizations across deployments, and so on.

QRadar Master Console helps to keep data (events and flows) segregated and shows only the offenses and managed hosts' status.

More on QRadar Master Console can be found at `https://www.ibm.com/docs/en/qsip/7.4?topic=monitoring-qradar-master-console`.

Query 4

Can you manage multiple customers' data (which should be segregated) in a single QRadar deployment – that is, a managed security service provider (MSSP)?

Yes. In IBM QRadar, you can define domains and tenants. Different customers can have data in the same QRadar deployment. The users who log in to QRadar UI for one customer will not be able to run searches for data belonging to other customers. This also means that users defined for one customer will not be able to view offenses or work on the offenses that belong to other customers.

This is achieved by performing domain management. When you define a domain, you can define the event collectors collecting events for that domain. You can also define the log source group that belongs to that domain. A domain will have tenant clients that are defined for it, as well as defined users.

You will find **Domain management** and **Tenant management** buttons on the **Admin** tab for QRadar. If you have a multi-tenant environment, my recommendation would be to start defining the network hierarchy and then the domain. Start by adding one log source each, and see whether all the services such as searches and offenses are segregated. Once set, you may then add the log sources in bulk to the log source group.

A detailed document on tenant management can be found at `https://www.ibm.com/docs/en/qradar-on-cloud?topic=administration-multitenant-management`.

Query 5

What do you mean by QRadar on Cloud or QRoC?

QRoC is the term often used for QRadar services provided by IBM. Depending on the amount of data to be processed and stored, IBM hosts the QRadar Console and/or managed hosts. IBM manages QRadar in terms of upgrades, upkeep, monitoring, and so on. Data gateways are installed on a customer site to collect data and send it securely to the IBM QRadar deployment in the cloud. Customers have access to QRadar GUI, with which they can search, check offenses, and create reports.

Detailed information on QRoC can be found here: `https://www.ibm.com/support/pages/getting-started-qroc`.

Query 6

When QRadar collects data, some of the data may be sensitive and should not be exposed even to the SOC analysts. How does QRadar comply with such a request?

There are scenarios where data such as credit cards or social security numbers should not be exposed to anyone performing a search in QRadar. This is known as data obfuscation, and this can be achieved by creating a profile and defining what kind of data should be hidden. Regular expressions are used to define the data that needs to be hidden.

Query 7

Is it possible to maintain the integrity of data collected and processed by QRadar?

The data collected and processed by QRadar should not be tampered with. If there are any changes made to the data, the integrity is lost. To ensure that data is not tampered with, QRadar can create a hash at the time the data is processed. If, later, the data is edited or tampered with, the new hash value will not match the older hash value and throw integrity errors. Hashing all the incoming data may not be ideal, as it will increase the CPU cycles and can cause performance issues.

Query 8

What are the different types of user authentication mechanisms that can be used in IBM QRadar?

By default, authentication is supported on the local system, and in the QRadar GUI, there is a **User Management** button, with which you can create new users and assign them roles. Logging in using these users is known as local authentication.

As QRadar users could be enterprise-wide, QRadar supports authentication using the following methods:

- RADIUS authentication
- TACAS authentication
- Active Directory authentication
- LDAP authentication
- SAML single sign-on authentication

Query 9

How do you go about sizing QRadar?

Sizing QRadar requires a lot of planning. The first thing to consider is the incoming event or flow capacity, which is measured in **events per second** (**eps**) and **flows per minute** (**fpm**). Then, we should consider the amount of time or the retention policy that a customer wants to configure. Depending on these factors, you should look at the **System requirements** page in the QRadar documentation. You will have a slight difference between system requirements when it comes to QRadar hardware and QRadar software installations. On the system requirements page, you will see details about the following:

- **Memory**: Always choose the recommended/suggested values instead of the minimum values in the chart. This will help you in the long run.
- **CPU**: Against each type of appliance, choose recommended/suggested or more CPUs when sizing.
- **Storage**: Always go with the supported storage options. Storage can also be expanded by using data nodes, coalescing events, configuring data retention, and configuring offboard storage.

More about offboard storage options can be found here: `https://www.ibm.com/docs/en/qsip/7.5?topic=storage-overview`

Along with the preceding factors, you should also consider configurational aspects – for example, an Event Processor that has data nodes attached, and you plan to store all the data in data nodes and none in Event Processor. In such a case, you can opt for less storage in Event Processor. Event Processor can be configured in "processing-only" mode when data nodes are attached.

> **Note**
>
> Minimum hardware requirements for installing QRadar on virtual appliances can be found at `https://www.ibm.com/docs/en/qsip/7.5?topic=installations-requirements`.

Query 10

How can you migrate the QRadar Console or managed hosts to new hardware if there is a disaster?

QRadar has a Data Synchronization app that can be installed on QRadar. This can help to sync data and configurations across your two sites. If there is a disaster, you can move your operations from the disaster recovery site.

Note that the Data Synchronization app *only* supports QRadar deployments where you have one-to-one mapping of QRadar appliances in your production site and disaster recovery site.

Maintaining exact replicas of QRadar appliances may not always be the feasible option in terms of costs.

There is another way you can keep configuration backup on a third-party server that can later be used for disaster recovery. Here, QRadar hardware migration can be done for the Console or the managed hosts if they are affected.

When dealing with the Console, there are two options. One is where an IP address remains the same after migration, and the other is where we provide a new IP address for the new site.

All the technical details on how different types of migration are carried out can be found at `https://www.ibm.com/docs/en/qsip/7.5?topic=hardware-qradar-siem-migration-scenarios`.

Query 11

As Linux admin, what are a few things that QRadar admin should be aware of?

As Linux admin, we usually change the IP address of servers from the command line, using Linux tools such as `ifconfig` or some networking scripts.

However, with QRadar, you cannot just change an IP address. If you want to change the IP address of the QRadar Console and there are managed hosts added, then you first need to remove the managed hosts from deployment. Then, change the IP address of the QRadar Console using a QRadar script called `qchange_netsetup`.

Detailed steps can be found at `https://www.ibm.com/docs/en/qsip/7.5?topic=nsm-changing-network-settings-qradar-Console-in-multi-system-deployment`.

Similarly, you cannot change the hostname of the QRadar Console or managed hosts. With this, we come to the last section of the last chapter of this book. QRadar SIEM has been a solid security solution for thousands of customers for years, and it will continue to do so. However, at the same time, IBM has come up with a new platform on which multiple security products such as QRadar are pre-installed. In the next section, we will look at the next-generation QRadar solution.

A next-generation QRadar sneak peek

A university in Canada worked on how they could monitor packets flowing in and out of their college network and saw whether they could find any anomaly or detect any threat. Later, a company called Q1 Labs was formed, which added event data monitoring to the flow data that was captured. Q1 Labs grew rapidly and was then bought by IBM around 2013. From that point, it was called IBM QRadar.

The IBM research team has been working on a platform where many of the different security products can be integrated seamlessly to provide more value to customers. The platform for the integration of security products is called **Cloud Pak for Security**, also known as **CP4S**. IBM has created multiple Cloud Pak solutions for data, automation, and so on.

We know that IBM QRadar is based on Red Hat Enterprise Linux. Similarly, CP4S is based on Red Hat OpenShift technology. This helps us to install CP4S wherever Red Hat OpenShift can be installed, so CP4S can be installed on a public or private cloud or even on-premise hardware. CP4S thus supports a hybrid model too.

IBM also provides CP4S as a service, which means that IBM will manage the platform called CP4S for the customer, and they can integrate whatever security tools they have on this platform. They can bring in their EDR, NDR, and risk management tools and integrate all of them on the CP4S platform.

Now, let us understand what the different IBM security products that integrate seamlessly with CP4S are:

- **IBM QRadar SIEM**: This is definitely a must-have product, as it brings with it the intelligence of correlating tons of data to create specific security incidents. These incidents as well as data can be pulled by CP4S.

- **IBM QRadar SOAR**: If there is a SIEM that generates security incidents, we should also have a SOAR to manage and respond to these security incidents. Playbooks are written and automated using SOAR. There are numerous applications that can be installed on CP4S, out of which Case Manager is the application that takes care of incident responses in CP4S.

- **IBM Security Randori Recon**: Attackers or adversaries always try to do recon to understand an attack surface that is exposed. This product helps us reduce the attack surface by discovering risks associated with the assets. Randori helps us to understand the assets that are exposed, prioritizing the risks.

- **IBM Security ReaQta**: This is IBM's EDR solution IBM. ReaQta has agents for both Windows and Linux endpoints. It does not have signature-based detection, but it has a strong behavioral-based approach. It has a proprietary Nano OS, which goes undetected by adversaries. This Nano OS is capable of monitoring different running services, creating a baseline for them. Whenever behavior changes, alerts are sent. This also helps to prevent zero-day attacks. From a response point of view, ReaQta agents are capable of isolating machines based on the response.

- **QRadar NDR**: In previous chapters, we discussed how flows can never be manipulated, unlike event data. Monitoring flow data can help you unearth different types of attacks such as lateral movement and data exfiltration. QRadar components such as **QRadar Network Insights (QNI)** and **QRadar Incident Forensics (QRIF)** and apps such as Network Threat Analytics and DNS Analyzer are part of QRadar NDR. The flows can be stored on **QRadar Network Packet Capture (QNPC)** appliances.

  Even though events may help to generate alerts, flow data will help you enrich data for SOC analysts. It helps them to understand an attack in depth and find the different assets that may be impacted.

- **QRadar XDR Connect**: Every customer will have security solutions and tools from different vendors. It is a challenge to integrate all these tools. In response, IBM came up with QRadar XDR Connect, which helps to connect these disparate sources, perform case management and investigations, as well as run threat hunting using the Kestrel threat hunting language.

This section focused on different IBM Security products coming together to create a security suite, or integrating into a platform called CP4S. Examples of these include QRadar on Cloud, which provides SIEM services to customers without installing them in-house, and a cloud solution in CP4S called **CP4S as a Service (CP4SaaS)**. These are the new technologies to look forward to. However, QRadar SIEM will remain the backbone on which other technologies and products grow and flourish.

Summary

In this chapter, we covered many practical aspects, such as troubleshooting QRadar. We discussed broadly the different categories of issues seen in QRadar and how you should address them. Then, we looked at a selection of FAQs on QRadar. Finally, in the last section, we covered how QRadar shapes up and how it is now part of a larger solution suite.

As they say, with great power comes great responsibility, which also applies when using QRadar. As a QRadar administrator or analyst, there are tons of features, apps, and extensions that you can use, and if you know how to maintain and troubleshoot basic issues of QRadar while using them, you will go a long way. IBM provides support for QRadar 24x7 and will always guide you on all aspects of maintaining and troubleshooting QRadar.

Further reading

IBM, with its vast experience as security solution vendor, has come up with multiple resources for customers on how to use and troubleshoot their products. For IBM QRadar, there are the following:

- The IBM QRadar 101 page: `https://www.ibm.com/community/qradar/`

- IBM Fix Central: `https://www.ibm.com/support/fixcentral`

- IBM QRadar RFEs (now called IBM Ideas): `https://www.ibm.com/support/pages/qradar-requesting-new-features-ibm-ideas`

- IBM QRadar community/forum: `https://community.ibm.com/community/user/security/communities/community-home?CommunityKey=f9ea5420-0984-4345-ba7a-d93b4e2d4864`

- IBM X-Force Exchange portal: `https://exchange.xforce.ibmcloud.com/hub`

- IBM Learning Academy: `https://www.securitylearningacademy.com/`

- IBM Knowledge Center: `https://www.ibm.com/docs/en/qsip/7.5`

Index

Packtpub.com

Subscribe to our online digital library for full access to over 7,000 books and videos, as well as industry leading tools to help you plan your personal development and advance your career. For more information, please visit our website.

Why subscribe?

- Spend less time learning and more time coding with practical eBooks and Videos from over 4,000 industry professionals

- Improve your learning with Skill Plans built especially for you

- Get a free eBook or video every month

- Fully searchable for easy access to vital information

- Copy and paste, print, and bookmark content

Did you know that Packt offers eBook versions of every book published, with PDF and ePub files available? You can upgrade to the eBook version at packtpub.com and as a print book customer, you are entitled to a discount on the eBook copy. Get in touch with us at customercare@packtpub.com for more details.

At www.packtpub.com, you can also read a collection of free technical articles, sign up for a range of free newsletters, and receive exclusive discounts and offers on Packt books and eBooks.

Other Books You May Enjoy

If you enjoyed this book, you may be interested in these other books by Packt:

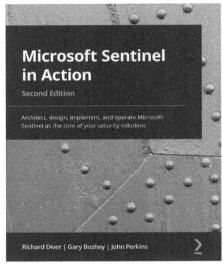

Microsoft Sentinel in Action - Second Edition

Richard Diver, Gary Bushey, John Perkins

ISBN: 978-1-80181-553-6

- Implement Log Analytics and enable Microsoft Sentinel and data ingestion from multiple sources
- Tackle Kusto Query Language (KQL) coding
- Discover how to carry out threat hunting activities in Microsoft Sentinel
- Connect Microsoft Sentinel to ServiceNow for automated ticketing
- Find out how to detect threats and create automated responses for immediate resolution
- Use triggers and actions with Microsoft Sentinel playbooks to perform automations

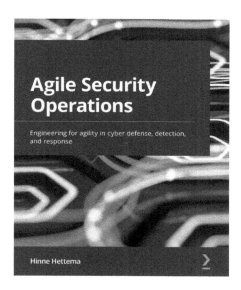

Agile Security Operations

Hinne Hettema

ISBN: 978-1-80181-551-2

- Get acquainted with the changing landscape of security operations
- Understand how to sense an attacker's motives and capabilities
- Grasp key concepts of the kill chain, the ATT framework, and the Cynefin framework
- Get to grips with designing and developing a defensible security architecture
- Explore detection and response engineering
- Overcome challenges in measuring the security posture
- Derive and communicate business values through security operations
- Discover ways to implement security as part of development and business operations

Packt is searching for authors like you

If you're interested in becoming an author for Packt, please visit `authors.packtpub.com` and apply today. We have worked with thousands of developers and tech professionals, just like you, to help them share their insight with the global tech community. You can make a general application, apply for a specific hot topic that we are recruiting an author for, or submit your own idea.

Share your thoughts

Now you've finished *Building a Next-Gen SOC with IBM QRadar*, we'd love to hear your thoughts! Scan the QR code below to go straight to the Amazon review page for this book and share your feedback or leave a review on the site that you purchased it from.

`https://packt.link/r/1801076022`

Your review is important to us and the tech community and will help us make sure we're delivering excellent quality content.

Download a free PDF copy of this book

Thanks for purchasing this book!

Do you like to read on the go but are unable to carry your print books everywhere?

Is your eBook purchase not compatible with the device of your choice?

Don't worry, now with every Packt book you get a DRM-free PDF version of that book at no cost.

Read anywhere, any place, on any device. Search, copy, and paste code from your favorite technical books directly into your application.

The perks don't stop there, you can get exclusive access to discounts, newsletters, and great free content in your inbox daily

Follow these simple steps to get the benefits:

1. Scan the QR code or visit the link below

https://packt.link/free-ebook/9781801076029

2. Submit your proof of purchase

3. That's it! We'll send your free PDF and other benefits to your email directly

www.ingramcontent.com/pod-product-compliance
Lightning Source LLC
Chambersburg PA
CBHW060128060326

40690CB00018B/3798